GENES AND GENDER
FIRST IN A SERIES
ON HEREDITARIANISM AND WOMEN

Gordian Science Series Number I

GENES AND GENDER: I

Genes and gender conference,
1st, New York, N.Y., 1977

Edited by
Ethel Tobach and Betty Rosoff

Graphics by
Betti Broadwater Haft

GORDIAN PRESS
NEW YORK
1978

GORDIAN PRESS, INC.
85 Tompkins Street
Staten Island, N.Y. 10304

First Edition

Copyright © 1978 by
Ethel Tobach and Betty Rosoff

Library of Congress Catalog Card Number 78-50640
ISBN 0-87752-215-4

The Editors and participants thank the American Musem of Natural History for the use of its facilities during the Conference, and the New York Academy of Sciences for its hospitality. The paper given at the Conference by Dorothy Burnham has appeared in *Freedomways*, and we thank them for agreeing to have an expanded version of it printed in this volume. The paper given by Helen Block Lewis will appear in *Bio-Ethics and Human Rights: A Reader for Health Professionals*, edited by Elsie L. Bandman, R. N., Ed. D., and Bertram Bandman, Ph. D., published by Little, Brown & Co., Boston. We also thank Random House, Inc., who make it possible for Joan Probber and Lee Ehrman to use the figure showing abnormal human karyotypes which was the original work of Dr. William Loughman and appeared in *Biology Today*. Finally, we are grateful to Kate Flores whose efforts brought about the realization of our plan to publish the papers presented at the conference.

PROLOGUE

The recent publication of E.O. Wilson's *sociobiology* and the previous awarding of the Nobel prize to Konrad Lorenz and Niko Tinbergen have strengthened the "scientific" legitimacy of "hereditarianism." By hereditarianism we mean the dogma that genes determine an individual's life history in the most significant ways. In other words, each of us has a "genetic destiny" that programs our behavior according to race and sex. Defenses of sexism and racism in the name of evolutionary theory have been used to support the continuing attacks on the few victories won by women in the United States, such as antiabortion legislation, ERA defeats, and legal actions against affirmative employment programs. These events and the attempts to pit women against Blacks, Hispanics, and other minorities in a period of increasing unemployment have made it clear that it is necessary to expose the myth of genetic destiny. That myth says that women are doomed to exploitation because their genes determine their anatomy, physiology, and behavior. This then limits their societal activity and prevents them from overcoming their oppression.

Accordingly, the Genes and Gender Conference was organized by members of the Metropolitan Chapter of the Association for Women in Science, the Committee for Women in the American Museum of Natural History, the Ad Hoc Committee for Women in Science in the New York Academy of Sciences, and local members of the Division for the Psychology of Women of the American Psychological Association. The meeting, which is to be the first of a series, took place on January 29, 1977, when the temperature was 2° F. We were sure that ordering coffee for 50 people would be more than sufficient. More than 350 people showed up; parents with children, gray-heads, working women, people of many ethnic backgrounds, academics and students. WBAI recorded a lunchtime conversation with the speakers

7

which has now been broadcast several times. Small discussion groups grappled with the problems presented by the speakers.

The $3 registration fee designated to handle the expenses (which amounted to $150 and was based on an attendance of 50 people) resulted in a nest egg of $300. This might have been greater but we stopped collecting money from those who found it difficult to contribute. At the concluding session, the participants voted not to take their registration fees back but to use the money for publishing the proceedings, and so here they are.

We are grateful that we can write a few words at the end of these proceedings. We are sure you will find the papers stimulating. We thank all those who prepared papers for the Conference and publication, and particularly Florence Brauner and Ruth Manoff for their editing, and Betti Haft for the graphics. In addition, the following group leaders and expediters were instrumental in making all this possible:

Dori Bates; Judith Bellin; Anne Carten; Ann Collins; Rachelle Fishman; Jane Harada; Audrey Haschemeyer; Frederica Leser; Linda Mantel; Constance Martin; Mary Moller; Agnes O'Connell; K. Joyce Prestwidge; Daphne Prior; Anita Pruzan; Reva Rubenstein; Susan Sacks; Nina Tolchin; Janet Weisberg; and Ann Welbourne.

Betty Rosoff and Ethel Tobach

TABLE OF CONTENTS

Biographical Sketches

Anne Briscoe, Ph.D., from Yale University in 1949; biochemist; Assistant Professor in the Department of Medicine at the College of Physicians and Surgeons and Director of the Biochemistry Laboratory at the Harlem Hospital Center of Columbia University; Immediate Past President of the Association for Women in Science.

Dorothy C. Burnham, M.A., from Brooklyn College, in 1960; biologist; Assistant Professor at Empire State College of State University of New York; research on Black Women in slavery.

Fredrica Y. Daly, Ph.D., from Cornell University, in 1956; humanistic clinical psychologist on the faculty of Empire State College with a limited private practice; professional activities with special but forgotten women (office charwomen who consider themselves "aliens" and adult women who are returning learners ["returnees"] in guiding them to careers in person-oriented fields [appropriate technology]).

Lee Ehrman, Ph.D., from Columbia University, in 1959; geneticist; Professor in the Division of Natural Sciences at the State University of New York at Purchase, New York; co-author of *The Genetics of Behavior.*

Eleanor Leacock, Ph.D., from Columbia University, in 1952; social anthropologist on the faculty of City College and the Graduate School of the City University of New York. Field work among Native Canadians; in Zambia; in Swiss and Italian villages; as well as in and around New York City in the areas of education, race relations, and mental health; in all these contexts concerned with the varied and changing role of women.

11

Helen Block Lewis, Ph. D., from Coumbia University, in 1936; clinical psychologist in private practice and Professor (Adjunct) of Psychology at Yale University; author of *PSYCHIC WAR BETWEEN MEN AND WOMEN*

Joan Probber, B.S., from Columbia University, 1964; Ph. D. candidate in the Animal Behavior-Biopsychology Program of the City University of New York; co-author with Lee Ehrman of an article on genetics and behavior in the *American Scientist.*

Betty Rosoff, Ph.D., from the City University of New York, in 1967; endocrinologist; Professor of Biology, Stern College of Yeshiva University; President of the Metropolitan New York Chapter of the Association for Women in Science, 1978.

Ethel Tobach, Ph.D., from New York University, 1957. comparative psychologist at the Department of Animal Behavior, The American Museum of Natural History; research on the evolution and development of social-emotional behavior; writing on role of science in societal processes leading to racism and sexism.

PERTINENT GENETICS FOR UNDERSTANDING GENDER

Joan Probber, B.S. and Lee Ehrman, Ph.D.
Hunter College and State University of New York at Purchase

What is a woman? How and by whom are the parameters of gender fixed? The initial definition of female given in the 1961 Oxford English Dictionary is "belonging to that sex which bears offspring." Reproductive functioning is viewed as primary and from this flows an inexhaustible series of extremes through time; mother/whore, Beatrice/ La Belle Dame Sans Merci, Virgin Mary/Earth Mother, and more *ad infinitum*. It is apparently procreativity that has spawned all the myth, mystery, and misinformation that surround female functioning and our concepts of woman.

From an evolutionary point of view, the advantages possessed by the two sexes for the production of succeeding generations are impressive. Each sex is uniquely specialized; the female for eggmaking, producing an immotile sex cell (egg) with stored nutrients for a developing embryo; the male, for insemination via the formation of motile sex cells (sperm). This division of reproductive labor was undoubtedly selected because of greater efficiency but most important, for the enhanced variability provided by the genetic pooling of two organisms instead of the doubling up and splitting of but a single one. Such pooling of genes even takes place in single-celled organisms like paramecia.

Perhaps the most significant accomplishment in the evolution of mammalian reproductive strategies was the ability of the female to retain the young in her uterus where development could proceed despite possible dangers in any external environment. Along with this a special organ evolved, the placenta, to feed and retain the young until more advanced stages of development. This was an important change from one of our phyletic predecessors, the reptiles. They produced great numbers of eggs of which relatively few hatched and still fewer survived. Producing only a few eggs but insuring their survival guaranteed a sometimes small but nonetheless continual gain.

Intrauterine development of the fetus in the mammal plus postnatal nursing afforded the young protection at their most vulnerable stages. This latter resulted in a socialization process that was valuable in that young animals lived intimately with mature ones for long intervals. Their behavior could therefore be affected significantly while offspring were growing and developing. Throughout mammalian evolution, a recurring theme is the importance of this early period. Romer (1958) emphasized this fact: "It is perhaps an exaggeration, but not too great a one, to say that our modern educational systems all stem back to the initiation of nursing by ancestral mammals."

Contemporary humanity is the endpoint of a long evolutionary history, as was realized by Darwin (1871). The first people evolved approximately 35,000 to 40,000 years ago. Their appearance was accompanied by a rapid expansion, diversification, and improvement of culture. They buried their dead together with flowers and implements carefully distributed around the corpse, so it is not unreasonable to assume that they believed in an afterlife and had some form of religion. This was *Homo sapiens* — modern humankind. The evolutionary trend was apparently the development of intellectual capacity — the single feature that makes our species unique. Morphological trends such as an increase in brain size from about 500 cc in *Australopithecus* to about 1400 cc in *H. sapiens* plus bipedalism, plus behavioral advances in communication and toolmaking are evidences of this evolutionary trend. In addition to these we possess: (1) advanced toolmaking; (2) elaborate cultural organization; (3) additional increases in brain complexities; (4) a childhood and adolescence extended in time to provide a longer period in which to assimilate cultural achievements, and (5) a degree of control over the environment superior to that of any other species.

Modern man, appearing some forty thousand years ago, was anatomically indistinguishable from ourselves. Similarly, modern woman seems to have remained anatomically unaltered since then. What has changed radically are the psychological and social characteristics of the role — the farrago of attitudes, ideas, and assumptions about women now extant.

To indulge in a bit of etymological play, the ambivalence about gender seems built into the very words that are used to identify sex. Female (feeble male?), Wo-man (woe to man?) may be semantic sporting but attitudes affect language and one wonders to what extent these gender identifications reflect the view of the female as an unfinished male. The incorporation of this attitude into Freudian dogma and its current legacy is perhaps best exemplified by the anecdote of the four-year-old girl who was being bathed with her baby brother. She looked down at herself and then at her sibling and asked "Why am I so plain and he so fancy?"

That this contemporary ambivalence has historical roots in our vocabulary is confirmed in Skeat's Etymological Dictionary (1967). The word "woman" is a phonetic alteration of the Anglo-Saxon *wifman*, literally wifeman, the word *man* being formerly applied to both sexes. The word female with French and Latin derivations is noted as originally *femell*, by confusion and interchangeable with male; and also from Middle English *female* changed into female as "an accommodated spelling to make it look more like *male*." Thus, the obfuscations persist from our earliest ideas and their embodiment in our languages.

All of our relevant myths — our ideas symbolized — support the view of male worship, uneasiness or overt fright of the female. One of our more potent fables, that of Eve created from Adam's rib, nourishes the idea of male primacy. But quite the opposite is true physiologically.

"The implication is that the process of sexual differentiation involves the suppression of the development of the female's neurobehavioral system rather than the enhancement of the males. That is, in the absence of the male gonadal hormones, sexual differentiation proceeds according to a female ground plan. It seems that in the mammal, the basic genotypically determined sexual disposition is female" (Thompson, 1975).

Money (1965 and later) in a series of studies on sexual identity and gender, said that "without androgen, nature's primary impulse is to make a female – morphologically speaking at least."

Although the human embryo is predisposed to develop as a female, its sex will be determined by the types of chromosomes borne by the sperm. Our species has twenty-three pairs of chromosomes, each of which carries thousands of genes that compose the genotype which develops and unfolds to become the visible organism, the phenotype. In humans twenty-two of the total number of chromosomes are autosomes; that is, these chromosomes are present in two copies in both sexes. The twenty-third pair are the unique sex chromosomes; the female has two of the X chromosomes; the male has one X and one Y chromosome.

The X chromosome was first described in 1891 from insect material as "a peculiar chromatin body" but its function was not known then or for a long time after. Although the female had two doses of all the genes on the X chromosome and the male only one, there seemed to be no significant differences in X-linked gene products in females and males. It was not until 1962 that a basis for dosage compensation emerged. This has been named the Lyon hypothesis after Mary F. Lyon who first presented detailed studies on the subject. She hypothesized that in the normal female, one X chromosome is inactivated in each somatic body cell early in embryonic development. Whether the maternal or paternal X is rendered inert is a matter of chance, but patches of cells with different X chromosomal lineages emerge. The deactivated X becomes a tightly packed mass usually found close to the inner surface of the nuclear membrane of somatic cells in normal female mammals. It is called a Barr body after its discoverer. Accordingly, the male and female each have only one operative X chromosome. Since the female may express the contents of different X-borne genes physiologically, she is a mosaic. The male, however, has only one X, his maternal chromosome, to express.

What are the implications of this double X chromosome dose

for the female of our species? At the molecular level the X chromosome is a much more information packed and therefore biochemically active chromosome than the Y which is much smaller and except for the male sex determination and fertility, inert. Recent work by McKusick and Ruddle (1977) cited more than one hundred genes in the human X chromosome map. The Y chromosome bears only two known genes, one controlling the TDF (Testis Determining Factor) and the other coding for the H-Y antigen. The H-Y antigen plays an important role in the expression of male physiological and morphological characteristics, specifically for the differentiation of the testis. "Further male differentiation is imposed on the embryo by the action of testicular hormones against the inherent tendency toward the female phenotype" (Wachtel, 1977). It is also known that the male sex, with an X and a Y chromosomal complement, is relatively unprotected from the expression of unpaired deleterious genes on the single X chromosome. Everything on the maternal X chromosome is likely to be expressed if not modified by genes on the autosomes. This is demonstrated by the much greater percentage of males afflicted by color-blindness and hemophilia, among other hereditary conditions.

Every cell in the normal human being will contain copies of the forty-six chromosomes that characterize our species except for the sex cells — the ova (eggs) produced by the ovaries and the spermatozoa (sperm) produced by the testes.

During the formation of the eggs and sperm (gametes), the chromosome number is halved, a process known as meiosis, leaving the egg with twenty-two autosomes and an X chromosome and the sperm with twenty-two autosomes and an X or a Y sex chromosome. Males produce four sperm from each specialized cell in the testes and X and Y bearing sperm occur in about equal number.

In ovaries, meiosis also produces four products but three of these become polar bodies while the fourth becomes the chromosome-containing nucleus of the developing egg. Females are born with their entire reproductive quotient already present and the limits are set for the number of eggs they can produce (about 400 during a reproductive life span). Males, however, continue to produce sperm from sexual maturation until senescence.

Differences in gametic size and function are extreme. The head of the sperm is about 3 to 5 microns long and 2 to 3 microns wide. The tail which propels it, is ten times the length of the head. Essentially, it is a pared-down packet of miraculously organized DNA (genetic material) with a launching and traveling apparatus to respond to and fertilize ova.

The human egg, in contrast, is one of the largest cells in the body —a sphere about 130 microns in diameter. It moves only passively when it is essentially shoved along taking a few days to progress from ovary to uterus. As noted by Hartl (1977), "All the sperm that give rise to all the people that ever lived could be carried in a teaspoon....the eggs, would require a small bucket." The egg is comparatively enormous because in addition to its complement of chromosomes, its cytoplasm provides the nourishment for and biochemical assistance to the developing embryo for translation of its chromosomes. Unlike the female gamete the function of the sperm as a cell is complete when the sperm contacts and penetrates the egg (having shed its tail) and sperm and egg fuse.

With fertilization of the egg, formation of the embryo begins, but it is not until six weeks afterward that sex chromosomes become operative. At this point the Y chromosome, if present, will cause the inner layer of the gonads to develop into testes. If, however, there are two X chromosomes, at about twelve weeks the outer layer of the gonads will elaborate into ovaries.

At birth we can usually distinguish the way an organism appears (its phenotype) from the organism's underlying genotype, the chromosomes contributed by parents. The phenotype is much less stable. Genes control the manner and range of reaction to environmental pressures: the norm of reaction. In considering the norm of reaction concept, it is most fruitful to take an epigenetic approach. The environment of the developing embryo interacts with the genotype, modifies its expression and is actively involved at each level of development before and after birth. This interaction commences at a molecular level with the fertilization of the egg by the sperm. Enzyme activities and changes in the intracellular environment are initiated. The egg provides cytoplasm for the newly formed zygote and a well-primed machinery for protein synthesis which initiates the primary series of cleavages and invaginations that produce the embryo, fetus, and finally, the neonate. Control mechanisms fostering maturation are encoded within the genes, in the internal and external environment of the DNA, and in the interrelations and interactions between them. At every level—molecular, cellular, and organismic—there are functional gene environment interactions and products.

To examine the expression of these interactions, we must consider at least two laws that serve as the foundation, both historically and functionally, of the science of genetics (Mendel's Laws). One concerns segregation. Genes are particulate units of inheritance consisting of alleles, two or more variant forms of the same basic gene (Aa). These alleles separate intact when sex cells are produced bringing alternate forms of the same gene to a gamete (A or a). The second law deals with independent assortment. When the chromosomes separate to form eggs and sperm these variant alleles (AaBb) are transmitted independently of one another from one generation to the next. These alleles have various forms of expression, for example that of one blood type (A, B, O). When one allele is always expressed phenotypically in the organsm bearing it, it is called the dominant allele (capital A); a recessive allele is only expressed when the dominant allele is absent (small a). A dominant trait is expressed when one or two dominant alleles are present (AA or Aa) but one can only see the recessive phenotype when both recessives (aa) are present. Some alleles are not expressed as either dominant or recessive but are intermediate, and they produce a phenotype somewhere between the phenotypes of the parents (red, pink, white). Suppose that two individuals differing

FIGURE 1
TRI-HYBRID CROSS

SPERM \ EGGS	ABC	ABc	AbC	aBC	abC	aBc	Abc	abc
ABC	AABBCC	AABBCc	AABbCC	AaBBCC	AaBbCC	AaBBCc	AABbCc	AaBbCc
ABc	AABBCc	AABBcc	AABbCc	AaBBCc	AaBbCc	AaBBcc	AABbcc	AaBbcc
AbC	AABbCC	AABbCc	AAbbCC	AaBbCC	AabbCC	AaBbCc	AAbbCc	AabbCc
aBC	AaBBCC	AaBBCc	AaBbCC	aaBBCC	aaBbCC	aaBBCc	AaBbCc	aaBbCc
abC	AaBbCC	AaBbCc	AabbCC	aaBbCC	aabbCC	aaBbCc	AabbCc	aabbCc
aBc	AaBBCc	AaBBcc	AaBbCc	aaBBCc	aaBbCc	aaBBcc	AaBbcc	aaBbcc
Abc	AABbCc	AABbcc	AAbbCc	AaBbCc	AabbCc	AaBbcc	AAbbcc	Aabbcc
abc	AaBbCc	AaBbcc	AabbCc	aaBbCc	aabbCc	aaBbcc	Aabbcc	aabbcc

A TRI-HYBRID CROSS: EACH CELL REPRESENTS A DIFFERENT GENOTYPE. (SEE n=3 ROW IN TABLE 1)

20

in genes at only three different loci (genetically active sites on chromosomes) are crossed: AABBCC × aabbcc. The sex cells (gametes) are carrying ABC or abc while the first filial generation (F_1), or offspring, is entirely AaBbCc. Here, the possible gametes are greater in number, eight in fact: ABC, ABc, AbC, aBC, Abc, aBc, abC, and abc. This is true for each second generation (F_1) AaBbCc parent. Figure 1 indicates the number of possible genotypes when three genes are involved. For an idea of what occurs when more than three genes are involved, please see table 1. Since each chromosome has many thousands of genes the number of possible combinations is astronomical.

One sees that because sexually reproducing parents are known to have two different alleles (be heterozygous) for large numbers of genes, the possible combinations of parental genotypes greatly outnumber the actually ever-realized genotypes. The probability of two siblings acquiring the same genetic endowment from both or even one of their parents is infinitely small. This probability decreases as more and more pairs of genes are involved. And the possibility of two unrelated persons having the same genotype is negligible. Each of us is and shall always be genetically wholly unique.

What do genes have to do with behavior? One way to attempt an answer to this question is to look at what happens when routine genetic and chromosomal processes break down. Most sexually reproducing organisms are formed from two haploid gametes, that is, gametes with one set of chromosomes. Therefore, they are diploid, or have two sets of chromosomes. But the occurrence of three and even four sets of chromosomes is common in plants. In fruit flies triploid (three sets) and tetraploid (four sets) females also occur. In addition partially diploid mosaics exist. A chromosomal mosaic is an individual with tissues containing different chromosomal complements. This is due to abnormal cell division early in embryology. In humans, complete triploidy is soon lethal, but individuals who are diploid/triploid mosaics may survive although they are physically and mentally defective.

Table 1. Gregor Mendel's Laws of Segregation and Independent Assortment as applied to numbers of differing pairs of genes. F_1 heterozygosity is assumed as is complete dominance.

GENES	F_1 GAMETES	$F_1 \times F_2$ ZYGOTES	F_2 GENOTYPES	HOMOZYGOUS F_2 GENOTYPES	HETEROZYGOUS F_2 GENOTYPES	F_2 PHENOTYPES
n	2^n	4^n	3^n	2^n	$3^n - 2^n$	2^n
1	2	4	3	2	1	2
2	4	16	9	4	5	4
3	8	64	27	8	19	8
4	16	256	81	16	65	16
5	32	1,024	243	32	211	32
10	1,024	1,084,576	59,049	1,024	58,025	1,024

In contrast to modifications of chromosome number, there are those involving chromosome breakage. These include four possible types: deletion, duplication, inversion, and translocation. Deletion is the sometimes lethal removal of a gene or sequence of genes. Death may occur because a biochemical sequence has been excised, resulting in an intolerable structural or functional gap in the organism.

Duplication of a gene may cause an imbalance of gene activity, reducing the viability of an organism. However, since some organisms can tolerate duplications of genetic material, these duplications might play an evolutionary role if part or all of one of the duplications mutated and functioned differently from the original. Indeed, this is one way evolutionary change is postulated to occur. Greater genetic variability gives rise (in minute quanta, to be sure) to physiologic and morphologic differences, which endow the organism with an enlarged behavioral repertoire so that it can more successfully cope with an ever changing environment.

An inversion occurs when a chromosome breaks in one or two places and the segment between the break (s) then rotates 180° leading to a reversal of gene order and a resulting change in biochemical product. A translocation occurs when two chromosomes from different pairs break simultaneously and exchange segments.

According to current knowledge, not all of these chromosomal and gene changes are equally important in their effects on behavior. To date inversions and chromosome number changes appear to form the main categories of importance.

Trisomy for one of the smallest human autosome is called trisomy21 (three instead of two twenty-first chromosomes), or Down syndrome (fig. 2). One in 600 to 700 newborns from all human populations is afflicted with this syndrome. It is characterized by congenitally retarded mental, motor, and sexual development and a number of physical signs. (Such people usually do not live long and they usually score low on intelligence tests [less than 20 to less than 65]). Behaviorally these

FIGURE 2
ABNORMAL HUMAN KARYOTYPES PREPARED FROM CHROMOSOMES IN WHITE BLOOD CELLS

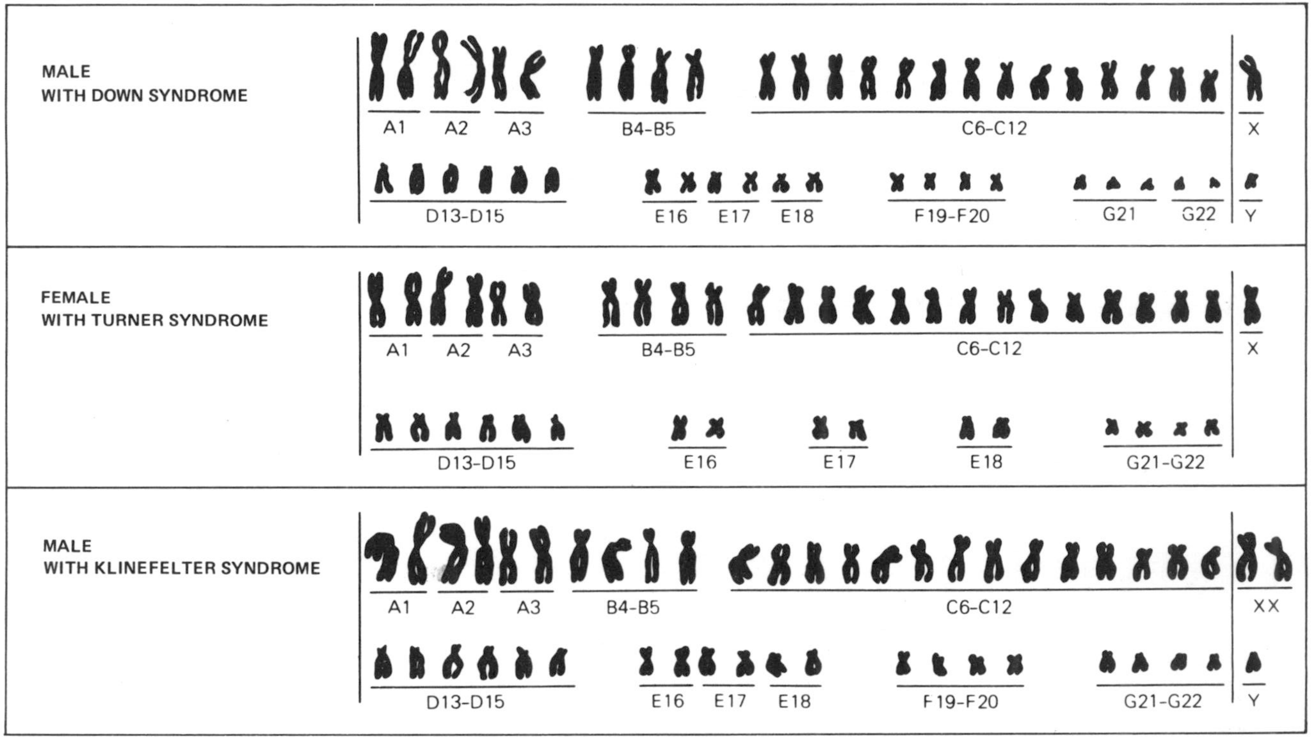

MALE
WITH DOWN SYNDROME

FEMALE
WITH TURNER SYNDROME

MALE
WITH KLINEFELTER SYNDROME

individuals are happy and friendly; they often imitate well, but inevitably do not show growth and development. Dingman (1968) studied psychological test patterns in patients with Down syndrome and other mentally retarded patients. He recorded no systematic differences in behavior.

The extra chromosome that characterizes Down syndrome presumably arises during the formation of the gamete when two of the chromosomes do not separate normally (nondisjunction). This is probably restricted primarily to the female gamete because the frequency of newborns affected by this increases rapidly with maternal age. For a woman 45 years old at pregnancy the risk of producing a child with Down syndrome is 1 in 50 compared with 1 in 3000 for a woman of 20 in the United States. Presumably, this increase in nondisjunction of the chromosome is due to a change in the environment of the ova brought on by increasing age.

It is generally accepted that autosomal aberrations appear to cause more severe effects (morphologically and behaviorally) than do X or Y chromosomal aberrations (fig. 2). For instance, Turner syndrome (gonadal dysgenesis) is characterized by absence of an X or a Y chromosome and therefore such people have 45 chromosomes instead of 46, written XO. In appearance those with Turner syndrome are female, though sterile; behaviorally they are characterized by normal verbal intelligence but visual infantilism in space perception—a specific space-form defect, i.e., a degree of space-form blindness or more technically, partial congenital agnosia (Schaffer, 1962; Money, 1970).

Klinefelter syndrome (bottom fig. 2) is caused by an extra X chromosome and therefore the person has 47 chromosomes, written XXY. In appearance these individuals are male and sterile. The syndrome is not necessarily marked by mental retardation, but it may be. XXY males are socially inadequate, often quitting school and other activities requiring social contacts. Some XXY males are more than socially inept; they are antisocial and may require institutionalization. Many reports list passivity, dependency, withdrawal from reality, limited interests, and poor impulse control as personality characteristics. The Klinefelter and Turner syndrome are among the products of sex chromosome aberrations and involve the addition or deletion of whole chromosomes with somatic effects that are rather minor compared with those of the Down syndrome (fig. 2, top). Turner syndrome occurs with a frequency of about 3 per 10,000 among newborns, and the frequency of Klinefelter syndrome is about 2 per 1000 newborns.

Although Turner women cannot bear children, their genetic and hormonal deficiencies do not adversely affect their heterosexual inclinations, their ability to marry, or their maternal interests.

In contrast to the lack of psychopathology in the Turner syndrome, Campbell and coauthors (1972) observed: "The incidence of psychiatric abnormalities in Klinefelter syndrome is far greater than in the general population." Examples of this are seizure disorders, speech disorders, electroencephalographic abnormalities, schizophrenia, paranoid states, and deviant sexual behavior.

And what of the notorious XYY human male? In 1967, Price and Whatmore filed the following report about a maximum security hospital in Scotland:

"The picture of the XYY males that emerges from examination of those detained at the State Hospital is of highly irresponsible and immature individuals whose waywardness causes concern at a very early age. It is generally evident that the family background is not responsible for their behavior. They soon come into conflict with the law, their criminal activities being aimed mainly against property, although they are capable of violence against persons if frustrated or antagonized. . .All nine men with an XYY chromosome complement conform fairly closely to this broad description and it seems reasonable to suggest that their antisocial behavior is due to the extra Y chromosome."

But is this a conclusion reached by too many too fast, as Levitan and Montagu (1977) and Witkin, Goodenough and Hirschorn

(1977) carefully caution? Novitski (1977) pointed out that surveys of newborns suggest that XYY may occur in as many as one in 1000 to 3000 live births in the United States and that this is much more frequent than the incidence of troublesome or even just subnormally intelligent tall men. In addition this high XYY frequency occurs in the absence of transmission of the chromosomal abnormality from father to son (Melnyk et al., 1969). Finally, men with the XYY karyotype do not exhibit antisocial behavior and many men who do exhibit antisocial behavior do not have this karyotype. Lawyers are therefore being advised to consider the evidence for and against this association in assessing legal responsibilities (Gardner and Neu, 1972).

Individuals with more than one X have sex-chromatin (Barr) bodies and are called sex-chromatin positive (female), whereas individuals with only one X are sex-chromatin negative (male).

Generally, the rule is:

Number of sex-chromatin (Barr) bodies =
number of X chromosomes - 1

The other rule about sex characteristics is that irrespective of the number of X chromosomes, the presence of a Y leads to a male phenotype (even if abnormal as in the case of the Klinefelter syndrome).

However, recent research (Wachtel et al., 1976) has concerned the phenomenon of a male phenotype without an integral Y chromosome. These researchers have demonstrated that the H-Y antigen (a male tissue recognition substance) is in fact made or controlled by a gene on the Y chromosome. This suggests the possibility that where there is no evidence of a Y chromosome but the individual looks like a male in every respect, the H-Y gene may have been translocated onto another chromosome in the course of development.

Since few genes are known to be borne by the Y chromosome, it is not surprising that XYY individuals occur without gross morphological abnormalities. Individuals with the Turner syndrome are female without a Barr body, while those with Klinefelter syndrome are male with a Barr body. In the case of

XXXY Klinefelter syndrome, two Barr bodies are expected and found. Because of the ease of staining a few cells in material scraped from the oral mucosa (the inside of the cheek) in which Barr bodies can be readily studied, such tests among other techniques, can provide important populational information on the frequency of "abnormal" males and females — at least those "abnormal" as regards sex chromosomes: and is not this precisely the sort of information currently needed?

It is beyond the scope of this paper to consider the genetic future of humankind further. We can predict that evolutionary changes will continue to occur and that evolutionary alterations in behavioral traits are likely to assume progressively more importance in the same way that selection for disease resistance must have been intense in communities before modern preventive medicine. Can we not then state with assurance that among these behavioral modifications will be crucial ones associated with our current standards of womanliness?

LITERATURE CITED

CAMPBELL, M., et al.
   1972. Klinefelter's syndrome in a three year old severely disturbed child. J. Autism Child. Schizo., vol. 2, pp. 34-48.

DARWIN, CHARLES
   1871. The descent of man and selection in relation to sex. London, J. Murray.

DINGMAN, H.
   1968. Psychological test patterns in Down's syndrome. *In* Vandenberg, S. (ed.), Progress in human behavior genetics. Baltimore, Johns Hopkins University Press, pp. 19-25.

EHRMAN, LEE, and P. PARSONS
   1976. The genetics of behavior. Sunderland, Mass., Sinauer Press.

GARDNER, L., and R. NEU
1972. Evidence linking an extra Y chromosome to sociopathic behavior. Arch. Gen. Psychiatry, vol. 26, pp. 220-222.
HARTL, DANIEL L.
1977. Our uncertain heritage: genetics and human diversity. Philadelphia, Lippincott.
KESSLER, S., and W. MCKENNA
1978. Gender: an ethnomethodological approach. New York, Wiley - Interscience.
LEVITAN, M., and A. MONTAGU
1977. Textbook of human genetics. 2nd edition. New York, Oxford University Press.
LYON, M. F.
1962. Sex chromatin and gene action in the mammalian X-chrosome. Amer. J. Human Genet., vol. 14, pp. 135-148.
MCKUSICK, VICTOR, and FRANK RUDDLE
1977. The status of the gene map of the human chromosome. Science, vol. 196, pp. 390-405.
MONEY, J. (Ed.)
1965. Sex research, new developments. New York, Holt, Rinehart and Winston.
1970. Behavior genetics: principles, methods and examples for XO, XXY, and XYY syndromes. Sem. Psychiatry, vol. 2., pp. 11-29.
MONEY, J., and S. MITTENTHAL
1970. Lack of personality pathology in Turner's syndrome: relation to cytogenetics, hormones and physique. Behav. Genet., vol. 1, pp. 43-56.
NOVITSKI, EDWARD
1977. Human genetics. New York, MacMillan.
OXFORD ENGLISH DICTIONARY
1961. A new English dictionary on historical principles. Oxford. Clarendon Press.
PRICE, W., and P. WHATMORE
1967. Criminal behavior and the XYY male. Nature, vol. 213, pp. 815-816.

ROMER, A. S.
   1958. Phylogeny and behavior with special reference to vertebrate evolution. *In* Roe, A. and G.G. Simpson (eds.) Behavior and Evolution, New Haven, Yale University Press.

SCHAFFER, J.
   1962. A specific cognition deficit observed in gonadal aplasia (Turner's Syndrome). J. Clin. Psychol., vol. 18, p. 403.

SKEAT, WALTER
   1967. A concise etymological dictionary of the English language. Oxford University Press.

THOMPSON, RICHARD F.
   1975. Introduction to physiological psychology. New York, Harper and Row.

WACHTEL, STEVEN, et al.
   1976. Serologic detection of a Y-linked gene in XX males and XX true hermaphrodites. New England J. Med., vol. 295, pp. 750-754.

WACHTEL, STEVEN
   1977. H-Y antigen and the genetics of sex determination. Science, vol. 198, pp. 797-799.

HORMONES AND GENDER

Anne M. Briscoe, Ph.D.
Assistant Professor of Medicine
Harlem Hospital Center
College of Physicians and Surgeons
Columbia University

This discussion of the relation of hormones and gender is prepared for the educated nonspecialist in biological sciences, for concerned feminists, and for those whose knowledge of endocrinology needs updating, as mine did when I delved into this. It represents an attempt to compare popular beliefs with the facts concerning a complicated and as yet incompletely researched aspect of endocrinology. The thesis is that the endocrine nature of the human species is hermaphroditic. That is to say, not only do both sexes possess identical metabolic hormones, but also the hormones concerned with reproductive physiology and gender characteristics are the same in both sexes. The human (and mammalian) hormone composition is a blend; sex differences in that mixture of hormones are quantitative but not qualitative.

The French say *vive la diffe'rence,* facetiously applauding the differences between the sexes. Unfortunately, *la diffe'rence* has lost something in translation, and has been interpreted as implying an unfavorable comparison with men in the intrinsic value of the female and in her value to society. In order to use biology to support the premise of the inferiority of women to men, the physiological attributes common to the human species have been ignored or minimized and the differences between the sexes have been exaggerated and misrepresented. In this discussion, we will attempt to identify the absolute and the relative as well as the quantitative differences from the viewpoint of modern endocrinology.

Endocrinology is the study of the glands or tissues that manufacture chemical compounds (hormones) with special physiological activity and which release or secrete these agents into the circulating blood to influence cells at a distance from the cells of origin. This release is not haphazard but occurs in response to specific signals and is similarly turned off by specific mechanisms. The chemical identity of the hormones, their physiological actions, and the factors that trigger or inhibit their release are all phenomena of concern to the endocrinologist. The cellular mechanisms by which these blood-borne messengers bring about physiological changes are still obscure, but the type and degree of response appears to depend on the presence in cells of specific hormone receptors and on the inherent capacity of the cell to respond. Thus different cells, such as those of the ovary and the testis, respond differently to the same hormonal agent.

The undisputed realm of distinct sexual difference between males and females is anatomical. Structural differences appear in fetal life and are maintained and extended into adult life. These differences include the gonads (i.e., the sex glands, the ovary, and testis) and the primary and secondary internal and external sexual structures. Moreover, all the cells of the entire organism are identifiable histologically as being part of a male or a female depending on the type of sex chromosomes their nuclei contain.

The questions relevant to the relationship between hormones and gender are: (1) Is there a qualitative difference in hormones between males and females?; (2) Is there a quantitative endocrine difference between the sexes?; (3) Do hormones create gender?; (4) Is there scientific evidence for a correlation between sex-related hormones and such attributes as intelligence, leadership, mathematical ability, critical judgment, and so forth?

The fourth question will be left to the behavioral scientists, except that I wish to point out that alterations in behavior, personality, or mental capacity have been clearly associated with several hormones that are not directly concerned with the physiology of sex. For example, irrational behavior is associated with severe hyperinsulinism or hyperthyroidism; retarded mental development is seen in severe hypothyroidism; and emotional depression is often a symptom of chronic hypoparathyroidism.

The first question is really whether or not the glands of internal secretion of males and females release the same hormones into the blood stream. The answer is affirmative. Those hormones not directly concerned with reproductive physiology are those manufactured and released by the various endocrine glands, i.e., pituitary, thymus, parathyroid, thyroid glands and by the endocrine tissues of the gastrointestinal tract, pancreas, kidney, adrenal medulla, adrenal cortex, hypothalamus, and the reticuloendothelium. Collectively, they produce a large number of hormones that are in all cases qualitatively identical in the sexes. Those hormones that are concerned with sexual phenomena are also qualitatively the same in both sexes. They are produced and released by the following scheme: the pineal gland, which regulates the hypothalamus; the hypothalamus secretes hormones that regulate the rate of pituitary hormone secretion; the anterior lobe of the pituitary gland secretes hormones that control the gonads and the mammary glands; the gonads in turn secrete sex hormones in response to stimulation by pituitary hormones called gonadotropins; the adrenal cortex or cortices secrete sex hormones in response to stimulation by another of the pituitary hormones, ACTH or the adrenocorticotropic hormone. In addition, in females the Graafian follicle in the ovary periodically produces the hormone progesterone (which is also present in males as will be explained later), and the placenta during pregnancy produces pituitary-like hormones. These last-named compounds are called chorionic gonadotropins and resemble the gonadotropic hormones secreted by the pituitary gland. The gonadotropins are identical in males and females and are essential for the health and normal activity of the gonads. The pituitary also secretes a hormone called prolactin in both sexes which in females functions in the post-partum state to initiate and maintain lactation. Beyond this, although its concentration in blood is increased in males in stress situations, its function is, thus far, less well defined. It appears to affect the growth and function of the prostate gland and to play a role in puberty in both sexes.

In elementary biology, we learned that the life cycle of the individual repeats the development of the species or "ontogeny recapitulates phylogeny." Thus, following the evolution of the species, the human embryo or fetus passes through a hermaphrodite state before it develops separate sexual anatomical structures. However, the endocrine system persists through life without qualitative differentiation in its secretions.

Gender is not derived from hormones. It is conferred on the new individual at the instant of fertilization when the nucleus of a spermatozoon fuses with the nucleus of an ovum and a one-celled organism results. This cell contains all the programming needed to direct the development of a female or a male person. In all embryos, a primitive sex gland develops in the first few weeks and is at first indistinguishable sexually. Later, in the individual with a Y chromosome, the primitive sex gland develops into a testis, and secretes one or more hormones. Presumably because of the action of these hormones, the sexually unique structures of the male duct system develop. On the other hand, the lack of the Y chromosome (and the hormone secretion it evokes), results in the development of the sexual anatomy of a female. Hence, the differentiation of the sexes in fetal life depends first on the presence or absence of a Y chromosome, and second on the presence or absence of a hormone produced by the fetal testis. After birth, as we have said, pituitary gonadotropins, which are identical in both sexes, regulate the activity of the gonads. They induce male or female reproductive development and function depending on the genetic origin of the tissues upon which they act. In females, the ovary responds by releasing sex hormones called estrogens, which induce female characteristics and participate in ovulation. In males, the testis responds by producing sex hormones called androgens which bring about growth and development of the testis, production of spermatozoa, and male characteristics. Thus the effect of the pituitary hormones is to maintain the secretion of the ovarian and testicular hormones in circulating blood. The sequence of events is as follows: the pituitary gland is stimulated to secrete its gonadotropins by releasing hormones reaching it from the hypothalamus. (The production of releasing hormones is inhibited before puberty by the hormone of the pineal gland, melatonin). The ovary and testis in response to the gonadotropins then increase their secretory rates of sex hormones in response to the pituitary gonadotropins. When the blood level of ovarian hormones rises sufficiently, the hypothalamus turns itself off, and then in turn, the pituitary gland becomes less active. This is followed by a reduction in output of hormones by the gonads. When the blood level of testicular or ovarian hormones declines, the hypothalamus is stimulated and in turn restimulates the pituitary gland. This endocrine homeostasis is set at a higher level of secretion in puberty to facilitate the growth of secondary sex characteristics, spermatogenesis, and ovulation. For a schematic representation of these relationships, see figure 1. One of the biological rhythms in females is, of course manifest as the menstrual cycle. This ebb and flow of sex hormones has been erroneously believed to be uniquely characteristic of female physiology. Recent studies, however, indicate that males have a similar rhythm of sex hormone secretion. Figure 2 illustrates one such study in normal adult men (Doering et al., 1975). The concentration in blood of the hormone testosterone is plotted against time in days. The data show a considerable scatter. However, the lines clearly show the cyclic trend. The length of the cycles ranges from 8 to 30 days, with the majority averaging 20 to 22 days. This should put in perspective the accusation that women are unstable due to the periodicity of their sex hormone secretions.

The hormones produced by the ovary and the testis, as well as

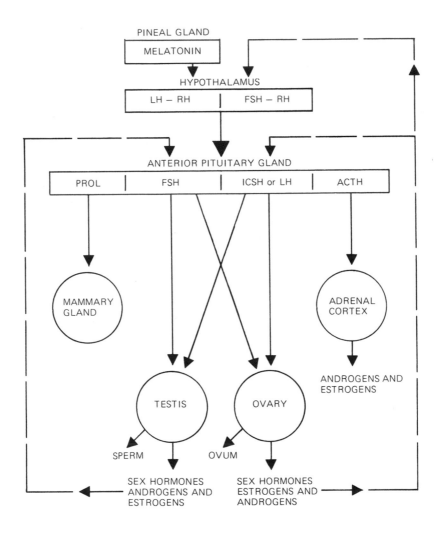

LH–RH = LUTEINIZING RELEASING HORMONE
FSH–RH = FOLLICLE STIMULATING RELEASING HORMONE
LH = LUTEINIZING HORMONE OF FEMALE WHICH IS IDENTICAL TO
ICSH = INTERSTITIAL CELL STIMULATING HORMONE IN MALE
PROL = PROLACTIN
ACTH = ADRENOCORTICOTROPIC HORMONE

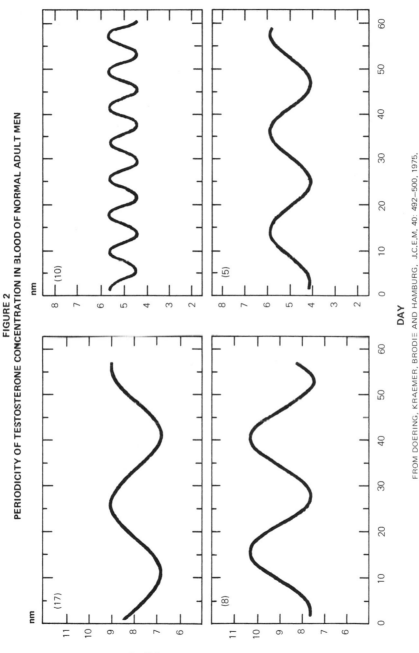

FIGURE 2

PERIODICITY OF TESTOSTERONE CONCENTRATION IN BLOOD OF NORMAL ADULT MEN

PLASMA TESTOSTERONE (ng/ml)

DAY

FROM DOERING, KRAEMER, BRODIE AND HAMBURG, J.C.E.M. 40: 492–500, 1975.

37

FIGURE 3
SEX HORMONES PRODUCED BY THE GONADS AND THE ADRENAL CORTEX

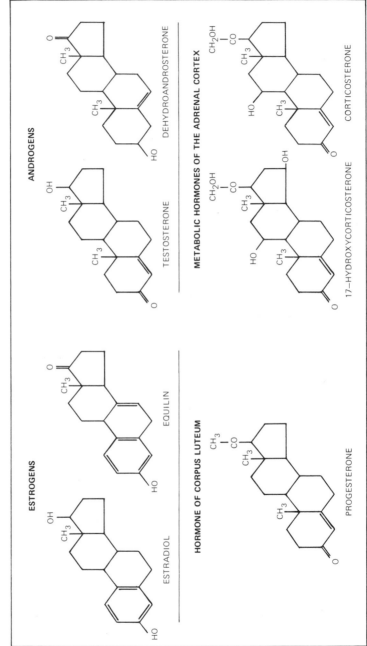

by the adrenal cortex, belong to a class of chemical compounds called steroids, examples of which are shown in figure 3. This is not intended to be mind-boggling to the social scientist or student of the humanities. One does not have to be a chemist to perceive the essential molecular similarity of these hormones. Of the vast number of compounds in living organisms, the only others that are chemically similar to these are cholesterol, the bile acids, and vitamin D.

The capacity of the adrenal cortex to produce hormones which are chemically similar to those of the gonads is explained by the common embryonic origin of these two pairs of glands. The primitive adrenal cortex develops in the embryo adjacent to the site of development of the gonads. Much later in the life cycle, the adrenal cortex, via its sex hormones, is needed for normal pubescence. That is, the normal maturation of a child into either an adult female or an adult male requires both gonadal as well as adrenal cortical sex steroids. Whether males require estrogens is less certain than that immature females require androgens as well as estrogens to achieve sexual maturity. Yet the word *androgen* comes from Greek word roots meaning "to produce a man" and the word *estrogen* comes from Greek word roots meaning "to beget mad desire" or frenzy. These terms, as well as the phrases "male sex hormone" or "female sex hormone" obscure the truth by implying an absolute difference between these compounds. The role of estrogens in the male does not seem to have been elucidated, but estrogens may, in fact, have a function related to inhibition of the hypothalamus. For example, the hypothalamic releasing factor for one of the pituitary gonadotropins, the follicle stimulating hormone or FSH, declines in the circulation when estrogens rise in the blood in both males and females, but not in response to androgens. Perhaps, then, estrogens function in males to turn off the pituitary gland's secretion of FSH via inhibition of the hypothalamus.

The top lefthand side of figure 3 shows a variety of estrogens. These are produced by the ovary, testis, and adrenal cortex. The

Table 1

Blood Plasma Concentrations of Sex Hormones in Human Adults

HORMONE	FEMALE		MALE	
	nanogram/ml		nanogram/ml	
Estradiol	0.3	(2)	0.09 ± 0.05	(3)
	(27 in pregnancy)			
Estrone			0.12 ± 0.05	(3)
Progesterone	1 (follicular phase)	(4)	0.25 ± 0.05	(5)
	10 (luteal phase)	(4)		
17-OH Progesterone			0.94 ± 0.23	(5)
Testosterone	$0.42 + 0.18$	(6)	4.5 ± 3	(3)
	0.2 to 1.0	(3)	3.7 to 10.0	(7)

Legend: 1 nanogram equals one millionth of a milligram or 10^{-9} gram.
1 ml equals one milliliter or one thousandth of a liter.
Numbers in parentheses refer to those in the literature cited.

top righthand side shows some of the androgens that are also produced by the gonads and the adrenal cortex of both sexes. The hormones in the right of the lower part of the figure are not sex hormones; they are metabolic hormones of the adrenal cortex. Progesterone, shown at the bottom left, is chemically related to both the adrenal and gonadal hormones. Estradiol, the most powerful estrogen, differs from testosterone, the most powerful androgen, by having one less hydrogen atom at the third carbon atom (O instead of OH), and by the absence of a methyl group (CH₃) attached at the tenth carbon atom. Thus, the androgens as a group have 19 carbon atoms and the estrogens have 18.

Studies of the biosynthesis or manufacture of these sex steroids show that androgens and estrogens are interconvertible in the body and that all are present in both sexes in different amounts. These data are tabulated in table 1. The data are recent and are from several sources. As you see, they do not all agree as to the magnitude of the differences. It seems clear, however, that males produce more testosterone (more is not necessarily better) and less estrogens. The amounts of androgens circulating in the blood are considerable in both sexes. The daily urinary excretion of these hormones is presented in table 2.

One wonders what this has to do with virility or femininity. What blood level of androgen is the cutoff point above which one gets to be a junior executive and below which one is relegated to a secretarial position? Another pertinent question is whether the exclusive role of estrogens is to produce female characteristics. Some answers come from other species of animals. According to Astwood "animals of the genus Equus including the horse are remarkable estrogen factories. The pregnant mare excretes over 100 mg daily, a record exceeded only by the stallion who despite clear manifestations of virility, liberates into his environment more estrogen than any living creature." (Goodman & Gilman, 1970)

The hormone mainly associated with female sexual physiology, and according to popular chauvinist view, with female inferiority, is progesterone. It has profound effects on the

Table 2

URINARY EXCRETION OF SEX HORMONES BY HUMAN ADULTS

HORMONE	UNITS	FEMALES	MALES
Estrogens	micrograms/day	5 - 100 (7)	4 - 25 (7)
		15 - 100 (2)	5 - 30 (2)
		16 - 60 (6)	0 - 20 (6)
Testosterone	micrograms/day	20 - 57 (2)	30 - 200 (2)
		0 - 15 (6)	47 - 156 (6)
17-Ketosteroids	milligrams/day	4 - 23 (7)	4 - 25 (7)
		10 ± 5 (2)	15 ± 5 (2)
		4 - 15 (8)	7 - 25 (8)
Progesterone as pregnandiol	milligrams/day	1 (follicular phase) (7,8)	1 (7)
		2-8 (luteal phase) (7,8)	

Legend: 1 microgram equals one millionth of a gram or 10^{-6} gram.
1 milligram equals one thousandth of a gram or 10^{-3} gram.
Numbers in parentheses refer to those in the literature cited.

uterus, uterine tubes, vagina, breasts, pituitary, and hypothalamus and is thought, in large amounts, to elevate the basal body temperature (i.e., temperature in the resting state). Recent investigations have established that it is produced in the adrenal cortex and placenta as well as the corpus luteum, and possibly the testis. Males have circulating progesterone in amounts not unlike those of the pre-ovulatory stage of the menstrual cycle in females as recorded in table 1. Progesterone has 21 carbon atoms, as do the metabolic cortical steroids, but lacks the alcohol (or OH) group at carbon atom 21, as they have, and has a methyl (or CH_3) group instead. Thus, not unexpectedly, its effect on the metabolism of salt and water is similar to the retaining action of the metabolic adrenal hormones, and the effect of progesterone in this respect is a cause of discomfort in many women. It has been shown to be intermediary in the biosynthesis of both estrogens and androgens as well as a precursor of the metabolic adrenal cortical hormones. Its role in males may be only that of a precursor. Its presence is certain if its role is not.

The phenomenon of pre-menstrual tension has been used as a pretext for discrimination against women, and it is said that the symptoms are due to progesterone. Pre-menstrually, women are accused of being irrational and unstable and therefore unsuited for appointment to positions of responsibility and decision making. Whether pre-menstruation is due solely and inevitably to progesterone, whether environmental factors play a role, and to what extent, are questions that have been insufficiently studied. Logically, if the ideas (or myths) about progesterone were accepted, one could argue that women who do not ovulate (for instance those taking the Pill), or who have had an ovariectomy, and all those who are post-menopausal should be considered to be equal to males, sharing with them the same amounts of this chemical demon. One wonders if androgens in excess are the cause of irrational, antisocial or criminal behavior in males rather than the source of male "superiority." If androgens are the secret weapon of jet pilots and bank presidents, if androgens are a prerequisite for effective aggressive behavior, then it must be argued that women make especially good use of their lesser share. Women compete well in traditional arenas (the department store bargain counter, or protecting their children), and they would compete at the highest echelons of business or government if they had the same opportunities as men. There are outstanding women in Congress, in the universities, in the National Academy of Sciences, in the honor societies such as Phi Beta Kappa, Sigma Xi, and Alpha Omega Alpha, in the armed forces, flying planes, designing buildings, preaching sermons, running companies, and waging a terrific struggle against institutional sexism in every phase of our national life. The odds against which women struggle would seem formidable or insurmountable to many males despite their greater supply of androgens.

Libido in both sexes depends partly on the presence of androgens circulating in the blood. Estrogens have also been shown in animal experiments to increase interest in sexual activity. It is interesting that the male, who has more androgen than the female, has a more fragile libido. It is also interesting that the correlation between gender identity and blood hormone concentrations is not significant. Finally, removal of the testes does not seem to affect intelligence or creativity as was demonstrated years ago in the Russian Skopec sect and in the Italian opera *"castrati"* both of which groups practiced gonadectomy before puberty. Yet because of socialization processes begun in infancy, men consider the essence of themselves to be an inseparable combination of their "brains" (cerebral cortex) and their sex organs. The knowledge of the truth of fertilization, that the sexes contribute equally to the creation of the new individual, was learned relatively late in history (the end of the nineteenth century). Perhaps this aspect of the male supremacy myth was bred into world culture inadvertently by Aristotole's prestige. He gave his authority (unquestioned for centuries) to erroneous views of reproduction:

namely, that life was conferred by the male sex fluid alone and the female merely supplied the nourishing medium *after* life was created.

In the twentieth century, further scientific progress was made. Scientists demonstrated that the new individual at fertilization acquires half its chromosomes from the sperm and half from the ovum but all or nearly all of its cytoplasm from the ovum. Furthermore, it begins life as a potential female unless it has a Y chromosome which intervenes, and as a result, it is programmed to produce different concentrations of qualitatively identical endocrine secretions to those of the individual lacking a Y chromosome (females). Individuals with a Y chromosome (males) are destined to have greater problems of socialization, of sexual identity, and of survival to old age. Perhaps the female endocrine mixture is a superior brew. Perhaps it would help to identify the recipe as hermaphrodite.

Summary

1. The hormonal endowment in both sexes is qualitatively comparable, and there is no known hormone which is unique to either sex.
2. Gender is genetically conferred and is distinct in every cell of the body.
3. The potential sex of the embryo is female unless a fetal male hormone is produced to direct the development of male anatomy.
4. Anatomically, the sexes become distinct in fetal life but a common endocrine system persists throughout life.
5. Males and females are a blend of androgens and estrogens. In this sense, the endocrinology of sex is not separate but hermaphrodite in the human species.
6. The pattern of secretion of sex hormones in both sexes is cyclical.
7. The progesterone concentration in circulating blood is similar in males and non-ovulating females (in the pre-ovulatory and postmenopausal states as well as in those taking the Pill).

GLOSSARY OF TERMS

Adrenal Cortex—Part of the adrenal gland. The adrenal glands are two small, yellow triangular glands situated above the kidney. Each one consists of two parts, the covering or cortex and the medulla inside; two parts with unrelated functions.

Androgen—A hormone which tends to stimulate the development of male sex characteristics but which also contributes to normal development of females in adolescence. Ex., Testosterone.

Chauvinist—Short for male chauvinist. In modern usage, a person who opposes the feminist movement; an advocate of the cause of male supremacy in society and its institutions. An acquired characteristic.

Corpus Luteum—An endocrine gland that forms from the reorganized Graafian follicle in the ovary after ovulation. A source of the hormone, progesterone.

Estrogen—A hormone which stimulates the development of female sex characteristics. Ex., Estradiol.

Feminist—A person, female or male, dedicated to advocating the extension of economic, professional, and political equality to women.

Fetus—An unborn, developing human, usually implying the period from three months after conception to birth.

Embryo—An unborn vertebrate, including human, at any stage from conception to birth. Used interchangeably with *fetus*.

FSH—The follicle stimulating hormone produced by the anterior pituitary gland. Identical in males and females. Stimulates and nourishes the gonads. Essential for maturation of ovum and sperm.

Gonad—Denotes the primary sex glands: ovary in the female and testis in the male.

Gonadectomy—Removal of one or both ovaries or testes.

Gonadotropic Hormone—Also called gonadotropins. Hormones, manufactured and released by the anterior pituitary gland, which nourish and stimulate the gonads and without which infantilism and/or sterility results. Ex., FSH, LH, ICSH.

Graafian Follicle—A small, spherical structure in the ovary which contains the ovum.

Hermaphrodite—A person in which two opposing qualities are combined; literally, an individual having both female and male organs of reproduction.

Hormone—A chemical substance that is produced in special cells of a gland or organ and released into the blood. It circulates to other cells and tissues where it activates those having specific receptors for it and as a result, it alters their physiological activity. It may increase or decrease activity. Also called internal secretions.

Hyperinsulinism—Secretion of abnormally large amounts of insulin into the blood.

Hyperthyroidism—Secretion of abnormally large amounts of thyroid hormone into the blood.

Hypothyroidism—Secretion of abnormally small amounts of thyroid hormone into the blood.

Hypoparathyroidism—Too little production of parathyroid hormone by the parathyroid gland.

Hypothalamus—A subdivision of the brain functionally related to both the pituitary gland and the autonomic division of the nervous system.

Lactation—The secretion of milk by the mammary glands.

LH or ICSH—The luteinizing hormone or the interstitial cell stimulating hormone. An identical chemical substance in females and males. In females it is essential for ovulation. In males it is essential for spermatogenesis (formation of sperm). (See Testosterone)

Metabolism—Any one or all of the chemical changes which take place in the body as a whole or in cells; the sequence of changes through which a particular substance goes as it is used in the body.

Ontogeny—The life cycle of an individual organism or person from conception to death.

Organism—A word denoting an individual that is capable of carrying on all the life functions. Any individual plant or animal including human individual.

Parathyroid Gland—One of four glands situated two on either side of the thyroid gland. Source of the parathyroid hormone that regulates calcium metabolism in the body. Essential for life.

Phylogeny—The evolutionary history of one group of organisms; i.e., the evolutionary history of a subdivision of the animal kingdom or phylum.

Pineal Gland—A small gland near the base of the brain. Produces a hormone called melatonin which acts upon the hypothalamus and prevents the secretion of LH Releasing Hormone. Potential contraceptive and abortogenic substance.

Progesterone—Hormone produced by the corpus luteum, the gonads, and the adrenal cortex. An intermediate substance or precursor in the manufacture of the sex hormones and adrenal cortical hormones. Found in glands and blood of both sexes but in large amounts in females in luteal stage of menstrual cycle and in pregnancy. Essential for successful pregnancy.

Post-partum—Period immediately after birth. Adjective or noun.

Prolactin—Hormone produced by the anterior pituitary gland which stimulates the secretion of milk in the post-partum mother. Present in both sexes.

Pubescence—The processes of developing sexual maturity; the period of time during which these processes are occurring.

Releasing Hormones—Hormones produced in the hypothalamus which regulate the rate of release of tropic hormones by the pituitary gland, i.e., LH Releasing Hormone and FSH Releasing Hormone.

Steroid—A compound consisting of five rings of carbon atoms. Ex., Cholesterol, bile acids, sex hormones, adrenal cortical hormones, Vitamin D.

Testosterone—The most powerful of the androgens. Essential for spermatogenesis. Produced in both sexes by the gonads and adrenal cortex.

LITERATURE CITED

BEESON, P.B. and McDERMOTT,W.
1975. Textbook of medicine. Philadelphia, W. B. Saunders Co.
CHATTORAJ, SATI C., JOEL S. RANKIN, ADRIAN K. TURNER, and ERNEST W. LOWE
1976. Urinary progesterone as an index of ovulation and corpus luteal function. Jour. Clin. Endocrinol. Metab., vol. 43, pp. 1402-1405.
DOERING, CHARLES H., HELENA C. KRAEMER, H. BRODY, KEITH, and DAVID A. HAMBURG
1975. A cycle of plasma testosterone in the human male. Jour. Clin. Endocrinol. Metab., vol. 40, pp. 492-500.
GOODMAN, LEWIS S., and ALFRED GILMAN
1970. The Pharmacological Basis of Therapeutics. London, MacMillan Co., Fourth Edition.
GUTAI, JAMES P., WALTER J. MEYER, III, AVINOAM KOWARSKI, and CLAUDE J. MIGEON (5)
1977. Twenty-four hour integrated concentrations of progesterone, 17-hydroxyprogesterone and cortisol in normal male subjects. Jour. Clin. Endocrinol. Metab., vol. 44, pp. 116-120.
HARRISON T.R. (8)
1974. Principles of internal medicine. New York, McGraw-Hill.
NETTER, FRANK H. (2)
1965. The Ciba collections of medical illustration, volume 4: endocrine system and selected metabolic diseases. New York, Colorpress.

SPEROFF, LEON, and RAYMOND L. VANDE WEILE (4)
 1971. Regulation of the human menstrual cycle. Amer. Jour.
  Obstet. Gynecol., vol. 109, pp. 234-247.
VARMA, MADAM M., CAROL A. HUSEMAN, ANN J.
JOHNSON, and ROBERT L. BLIZZARD (3)
 1977. The effect of prolactin on adrenocortical and gonadal
  function in normal men. Jour. Clin. Endocrinol. Metab.,
  vol. 44, pp.760-762.
VALETTE, A., B. SERADOUR, and J. BOYER (6)
 1975. Plasma testosterone levels during the menstrual cycle.
  Jour. Clin. Endocrinol. Metab., vol. 40, p. 160.

Numbers in parentheses are references cited in Tables 1 and 2.

BIOLOGY AND GENDER

Dorothy Burnham, M.A.
Freedomways Associates
and
Metropolitan New York Learning Center
Empire State College

Recently there has appeared in the popular press and in the scientific journals a number of books and articles which set out to demonstrate that the gender of women is defined by human biology. Some of the most exciting research in modern biology concerns the molecules that determine heredity. And it is no wonder that James Watson was lyrical in his account of the joint effort that resulted in the publication of the structure of DNA. When persuading his sister to type the manuscript, he told her she was participating in perhaps the most famous event in biology since Darwin's books (Watson, 1968).

This elegant model of the gene is a product of the language, the instruments and the sophisticated technology of the mid-twentieth century. It appears to me to be somewhat irrational that the advanced ideas of genetics should be used by some scientists to support the antediluvian ideologies of racism and sexism.

The employment of prevailing authority to define the role of women in society is not, of course, new. For in determining that position, the establishment of the family structure and other aspects of the political and economic life of the culture depends. Whether the boundaries of women's "place in society" were erected with the bricks of theology or the cement of genetic determinism, the intention is that the barriers shall remain strong and sturdy.

Eliza Gamble (1916) writing a half century ago pointed out that it did not seem strange that theology should be used to validate man as an infinitely superior being but that she did expect with the dawn of a scientific age that the prejudiced doctrines would disappear. Ms. Gamble would no doubt be extremely surprised to read the recent headlines regarding the right of women to serve in all professions including the priesthood and find that the theologians are still trying to hold up the sky.

Through generations, both the literature of mythology and the literature of science have been in good measure addicted to trying to prove the superiority of the male. Natural scientists have described the obvious physical differences between men and women and have moved on from that point to concoct great theories about differences in intelligence, emotional make-up, and behavior. The scientists in an attempt to appear neutral declared that they were only investigating whether women are less objective, more nervous and less emotionally stable than men. By these means, responsibility is disavowed when the gratuitous representation of women is translated to mean women cannot handle decision-making positions.

Carroll and Charles Rosenberg (1973) cited nineteenth century physicians who wrote of the delicacy of the female nervous system as opposed to that of the human male. "Few if any questioned the assumption that in males the intellectual propensities of the brain dominated, while the female's nervous system and emotions prevailed over her conscious and emotional faculties." Physicians warned that the intellectual life if opened up to women might well deprive the species of good health. Some of the scientists of that period apparently believed that the brain and the reproductive organs drew on the same sources for nourishment and, therefore, feeding one meant depriving the other. And since the chief function of women was reproduction the choice was clear—*forego education and intellectual pursuits.* During this period, one of the factions among those who opposed the vote for women emphasized the genteel, gentle nature of women, claiming that their biology suited them more for the parlor and the kitchen than for political life. They spoke emotionally of the necessity of protecting women from the trauma of real life outside the home. This argument of course neglected mention of the majority of poor and working-class women whose labors kept them out of the parlors—if indeed they had parlors to be in. Sojourner Truth's stirring reminder of the position of Black women in America was a consciousness-raising experience for the women at the Akron convention of the Women's Suffrage movement (Rossi, 1974):

> Dat man ober dar say dat womin needs to be helped into carriages, and lifted ober ditches and to hab de best place everywhar. Nobody eber helps me into carriages, or ober mud-puddles, or gibs me any best place! And an't I a woman? . . . I have ploughed, and planted, and gathered into barns and no man could head me! And an't I a woman? I could work as much and eat as much as a man—when I could get it—and bear the lash as well! And an't I a woman? I have borne thirteen children, and seen' em mos' all sold off to slavery, and when I cried out with my mother's grief, none but Jesus heard me! And an't I a woman?

Women were denied the right to vote on the basis of a different biology but working-class women and Black women were forced to work 8-, 10-, 12- hour days in industry or in the fields and in addition bore the load of housekeeping and child rearing.

Of course similar biological theories were advanced to account for the obvious superiority of whites over Afro-Americans. The exploitation of slave labor and Blacks after slavery was also defended on scientific grounds. Blacks had smaller brains with less capacity. Blacks were inherently happy-go-lucky. Black women were born without morals. These were not areas in which the scientists had difficulty getting an ear. As soon as the theories were proposed, the mass media spread the ideas abroad.

Some of the old biological biases against women and Blacks and non-Aryans have disappeared only to be replaced with less primitive and more up-to-date trappings. Steven Goldberg titled his book *The Inevitability of Patriarchy: Why the Biological Difference Between Men and Women Always Produces Male Domination.* Margaret Mead (1973) reviewing the book for *Redbook* magazine extracts some astonishing quotations. "What is lacking in the male is an acceptance that radiates from all women except those few who are driven to deny their greatest source of strength. Perhaps." Goldberg says, "this female wisdom comes from resignation to the reality of male aggression; . . . while there are more brilliant men than brilliant women, there are more good women than good men."

There were contradictions between the 18th and 19th century beliefs about the biological heritage and destiny of women and the actual place women held in the society. Nowhere was this more evident than in the treatment of Black women slaves.

The slave was treated like a nonhuman in every respect. In addition to the physical cruelties inherent in the system, slave women were subjected to psychological and emotional traumas beyond belief. They were whipped for attempting to escape as well as for other infractions of the harsh slave laws as severely as were men. There are numerous descriptions of the beating of pregnant women and women who had just borne children. In the case of pregnant women, it was customary to dig a hole to accommodate her belly so that the developing fetus, the source of new income for the slave-holder, might be protected. Testimony of ex-slaves recorded in the Fisk University Studies and in the W.P.A. narratives document the fact that women bore the scars of the whips to their graves.

Aside from the cruel punishments, girls and women were deliberately scarred and branded and had teeth yanked out so that they might be recognized if they fled plantation drudgery. Newspapers of the day carried advertisments similar to this one from the *Natchez Courier* of August 24, 1838, "Ranaway a negro [sic] girl called Mary, has a small scar over her eye, a good many teeth missing, the letter A is branded on her cheek and forehead."

Common to the system was the constant breaking up of families and the separation of women from their husbands and their children. A major theme running through the slave narratives and the literature of anti-slavery societies is the witnessing of the heartbreaking partings. With profit only as the motive, owners never gave a second thought when selling children that were transported hundreds of miles away from mothers who would likely never again see them. The mystique that the instinctive nurturing qualities which supposedly differentiated women from men and fitted them only to the occupations of caring for the home and children, apparently did not apply to the Afro-American slave woman.

The language and customs of the slave-holding class made it clear to all that the Black woman slave was considered to be a nonperson. First of all she was clearly property and treated as such—tradable and taxable and at the disposal of the slaveholder in every aspect of her life. Like the farm animals, she was referred to as a breeder and the children born to her were *increase*—not sons and daughters. Slave ship records spoke of slaves as cargo, tonnage, pieces.

The slave woman faced daily the crucial problem of surviving and adapting to this life and at the same time maintaining the integrity of her personality and reaching out to encourage and succor other members of the slave community. It is a tribute to the quality of the human spirit that these women were able to endure and even prevail over the hardships inflicted upon them.

The thesis of Goldberg and others is that the male hormones function to produce aggressive behavior in males which leads to dominance and the rewards of leadership and superiority. I would like to note here that it has been my experience that, in most instances, Blacks and women are unrewarded for aggressive behavior and may very well be punished for exhibiting this characteristic.

E.O. Wilson (1975) extracted from his book *Sociobiology; The New Synthesis* for a *New York Times Magazine* article. His theory is that there may be a genetic base for altruism. "Human social evolution is obviously more cultural than genetic. The point is that the *underlying* emotion, powerfully manifested in virtually all human societies is what is considered to evolve through genes." Admittedly this is a more palatable idea on aggression than Goldberg's. If we must have a genetic basis for emotion, most of us socially minded people might prefer the genes to be coding for altruism rather than aggression. Wilson then proceeds to propose a genetic base for genetic behavior: "We must then add the qualities that are so distinctively ineluctably human that they can be safely classified as genetically based; . . . [among them] is the weaker but still strong tendency for sexually bonded women and men to divide their labor into specialized tasks."

I think that one needs stronger evidence for genetically based behavior than merely to say that's the way it's always been and therefore it must be inherent. (The unspoken conclusion is of course that that's the way it always will be.)

In an earlier day, some biologists looked into the microscope and saw in the spermatazoon a midget human waiting to be implanted in the uterus, where, after a proper period of gestation, it would be born to grow up, down and out. It was a more suitable explanation of the male's contribution to reproduction (expecially after the discovery of the sperm) than the Adam's rib story or the head of Zeus myth but it has since been replaced. It is possible that some of those who are now looking into the gene and seeing inheritance factors for intelligence, emotions and behavior may someday recognize that the interpretation is a little more intricate than they have made it appear.

Certainly among the most dangerous of the new hereditarian ideas are those that promote and encourage the sterilization of women. William Shockley (1974) has used his established reputation as a Nobel prize winner in physics to advance an organization called Foundation for Research and Education on Eugenics and Dysgenics. Basing his beliefs and statements on the research of Jensen and others, Shockley stated that Blacks are intellectually inferior and that this inferiority has a genetic base. He proposed therefore, ". . . to explore reaction and perhaps undertake a trial run on a voluntary sterilization bonus plan." To their credit the great majority of American geneticists and psychologists have disavowed this outrageous proposal and the faulty research upon which it is based.

However, the proposal has been made in a fertile climate and at a time when receptivity is at high tide. Reports from Puerto Rico, the Southern part of the United States, and from Native American groups confirm that minority women have been the victims of involuntary sterilization in alarmingly increasing numbers. And in fact, the daily papers have reported the sterilization of young Black women of 13, 14, and 15 years. In a period of worldwide starvation and hunger and epidemic poverty, sterilization can become an acceptable solution to social problems in part because it is justified by so-called scientific research. During the period of slavery in the United States, as I have said previously, there were biologists who wrote of the inherited stupidity, emotional instability, and amoral character of the slaves. Given the fact that eugenics is proposed as a way to improve the species, one wonders why sterilization was not proposed at that time. My intuition tells me that Shockley and Eysenck would have been lynched had they made such a proposal to the slave-breeders.

The conception of the biologically based inferiority of women and of selected nations or races led to such grave consequences as the oppression of women, the institution of slavery, the Fascist holocaust. Those in power have never hesitated to use whatever weapons come to hand to maintain and extend their power. And ideas are popular weapons. Particularly if people can be persuaded to adopt ideas that are against their own interests. Exploitation of divided peoples is a classic means of subjugation.

Those of us who have made a conscious effort to displant culturally implanted misconceptions about our fellow human beings know how difficult that is. Especially when everything in the culture deliberately or insidiously acts to reinforce the misconceptions. Scientists who are a part of our environment, then, not only are contributors to the pollution of ideas but are indeed the receivers as well. Cohen (1974), writing of the social functions of science remarked: "They link science with society in such a way that science amplifies the social signals which stimulates it, and even exaggerates the worst of them." If we examine the attitude of science and scientists toward women we find that we come out badly. For not only have the scientists helped to justify the subordinate position of women in the society, but the scientific establishment itself for the most part has not encouraged women scientists to become contributors or leaders. The contribution of Rosalind Franklin to the theory and development of the DNA model was an integral and essential part of the work. Yet she went largely unrecognized. And the Nobel Prize for Medicine and Physiology was awarded to Crick, Watson, and Wilkins in 1962 for their research and accomplishments. Watson acknowledged overlooking Franklin in the epilogue to his book *The Double Helix* saying, "and we both came to appreciate greatly her personal honesty and generosity, realizing years too late the struggles that the intelligent woman faces to be accepted by a scientific world which often regards women as mere diversions from serious thinking."

Many other women have suffered similar treatment from the scientific community but indeed the majority of women never reach the point of considering a scientific career because they are conditioned from grade school up to avoid mathematics and science. And because the hurdles are even greater for minority peoples, the Black woman's opportunities in science are infinitesimally small.

If I sound more like a Black feminist than a biologist, the reason is—that's what I was first. I truly believe racism and sexism interact and reinforce each other and the effort of both is not arithmetical but geometrical on the subject.

I am well aware of the issues raised by the scientific community in regard to freedom of inquiry and freedom of speech. I do not, however, believe that anyone in America today would be given research money, facilities or publication privileges to prove a thesis relating to some supposed inferiority of white Anglo Saxon males.

Scientists have faced the question of whether their research will be used to develop more advanced weaponry and some have made their choice. Scientists are facing the question of whether the gains to be made in recombinant DNA research is worth the hazard it poses to the world. They must also face the question of whether research that starts out with the premise that women are inherently inferior can yield anything of lasting value to the human community.

LITERATURE CITED

COHEN, ROBERT
  1971. Ethics and science. *In* Truitt, Willis, and T.W. Solomons (eds.), Science, technology and freedom. Boston, Houghton Mifflin, p. 141.
GAMBLE, ELIZA BURT
  1916. The sexes in science and history. New York, G. P. Putnam, p. vii.
MEAD, MARGARET
  1973 (Oct.). Review of "The inevitability of patriarchy," by S. Goldberg. Redbook, pp. 46, 47, 52.
ROSENBERG, C.S., AND C. ROSENBERG
  1973 (Sept.). The female animal; medical and biological views of woman and her role in nineteenth-century America. Jour. Amer. Hist., 1973, pp. 332-336.
ROSSI, ALICE. S. (ed.)
  1974. The feminist papers. New York, Bantam Books, p. 428.
SCHOCKLEY, WILLIAM
  1974 (Jan. 13). Letter to Station WWRL, New York. Reprint distributed by Foundation for Research and Education on Eugenics and Dysgenics, Sanford, Calif.
WATSON, JAMES
  1968. The double helix. New York, Atheneum, pp. 221, 226.
WILSON, E.O.
  1975 (Oct. 12). Human decency is animal. New York Times Mag., p. 38ff

COMMENTS ON "SCIENCE AND RACISM"

Frederica Y. Daly, Ph.D.
Metropolitan New York Learning Center
Empire State College

On the road of prejudice there are many hovels, chief of which are sexism, classism, and racism. Unfortunately, science and scientists are among their architects. John Lily in the "Mind of the Dolphin" asserted that it is impossible to understand research findings without knowing the belief and value systems of the author-experimenter, implying that we influence what we seek and find. Professor Burnham uncovers some of the ugly proclivities of our colleagues in her paper, "Science and Racism." She asserts that "the scientist's most original ideas are shaped by the culture, and the political and social milieu in which he or she operates." She indicts contemporary scientists who have allowed their personal beliefs to contaminate their professional work and who have misused the tools and principles of science to oppress especially the poor of minority peoples, citing the writings of Jensen, Shockley, Hutschneker, etc.

John Stuart Mill, a classic advocate of the rights of the individual, asserted in his excellent essay, "On the Subjection of Women" that the "subordination of one sex to the other is wrong, that it should be replaced by a principle of perfect equality, admitting no power or privilege on the one side, nor disability on the other." Professor Burnham reapplies this principle in her concise but vitally relevant article. She clearly pinpoints the disservices given by some workers of the helping professions whose uses of psychosurgery, behavior modification, and drugs have contributed more to vested institutional interests than to healing the hurt.

A black woman and a scientist, Professor Burnham has had to resist the constrictions of racism and sexism, a conjoined phenomenon described by Dr. Pauli Murray in her essay, "The Liberation of Black Women" and later labeled "double jeopardy" by Frances Beale in her essay by the same name. Both urge black women to fight for the right to be Black and Women, supporting the right to compete with men, black or white for jobs. Antedating both early in this century, Dr. DuBois in his powerful essay, "The Damnation of Women," wrote "the future woman must have a life work and economic independence. She must have knowledge. She must have the right of motherhood at her own discretion. The mincing horror at free womanhood must pass if we are ever to be rid of the bestiality of free manhood; not by guarding the weak in weakness do we gain strength, but by making weakness free and strong."

Elizabeth Morrell reporting in an article (Freedomways, vol. 15, no. 4, 1975) on the proceedings at the World Congress for the IWY, stated that the women delegates were not united against men but around issues that would benefit all peoples. It remains true that women in the sciences need to unite around the issues of sexism, racism, and classism.

It is important for readers to know that the background of Professor Burnham's paper, as well as the backgrounds of the other authors, is marked by the personal experiences of racism and sexism, which adds meaning to their professional evaluations. One wishes that both time and circumstance permitted her to expand her remarks beyond the limitations natural to such a conference.

PSYCHOLOGY AND GENDER
Helen Block Lewis, Ph.D.
Yale University

For the past six years I have been trying to understand the complicated interaction between an exploitative society and gender differences in human behavior. My thesis, briefly stated, is that our exploitative society injures the two sexes differently. It not only creates different distortions of gender identity in men and women, but also produces different forms of mental illness in the two sexes. These are some of the ideas I developed in my book, *Psychic War in Men and Women* (1976).

Let me put my thesis forward in a formula for understanding sex differences which has served me as a kind of shorthand statement of the important variables. Sex differences in human behavior (including gender identity differences) result from sex differences in chromosomal endowment (XX and XY on the 23rd pair of chromosomes) times the social inferiority of women (an offshoot of an exploitative social system) – and this interaction must be considered as the numerator over a common denominator: the affectionate or social, i.e., "cultural" nature of the human species, of both sexes.

I address the common denominator first. I base it on the accumulating evidence from anthropologists' studies of the differences between ourselves and nonhuman primates and on the relatively recent developments in the study of human infancy. Anthropologists, at least some of them, tell us that the most profound evolutionary change from nonhuman primates to ourselves is the appearance of human *culture* as our species' adaptation to its environment. And human culture is unique in the scope it gives to affectionate nurturing of the young during a prolonged childhood. As a result, the power of moral prescriptions, i.e., internalized affectionate bonds to govern human behavior from birth to death, is also unique to our species.

The emergence of morally prescriptive human culture has been understood as the outcome of the Great Apes' descent from the trees. As Elaine Morgan has shown in her delightful book, *The Descent of Women*, scientific speculation about the evolution of human culture has itself been dominated by androcentric, i.e., male-oriented, thinking. The image of Tarzan, the Mighty Hunter, coming down out of the trees has been the model against which human beings' upright posture, opposable thumb, tool-using. language, and advanced cerebral cortex have been described. Morgan suggests that human evolutionary advance was also governed by the survival needs of a hypothetical female (primate) ancestor trying to nurture her young during the millions of years of Pliocene drought. Morgan's thesis reminds us that human culture is an enormous evolutionary advance over primate society not only because the Mighty Hunter has grown mightier but because human culture allows so much greater scope for affectionateness as a force governing human behavior.

As one example, there is a great difference between nonhuman primates and ourselves in the amount of affectionateness that characterizes sexual behavior. Human sexual behavior is different from that of the primates in that sex life is no longer governed by the estrus cycle in the female, that is, by a built-in biological clock. Note again, that the significant evolutionary change has occurred in the female. With estrus absent, human beings have lost the prepotent hormonal stimulus to copulation which all other mammals have. But, in contrast, human sexual behavior is unique in that affectionate behavior—kissing, playfulness, touching—have all been incorporated into lovemaking. Among primates, affectionate behavior is quite separate from the act of copulation. Compared with primate sex, human sex is replete with revivals of affectionate intimacies which children shared with their parents. It is also unique in the incest taboo—a universal moral prohibition against sexual intercourse between parent and child.

It is worth a brief digression to remark on how, not only androcentric thinking, but also the Weltanschauung of an exploitative society influences our categories of thought about human culture. We are told that the human infant is the most helpless of all species: completely at the mercy of its parents who, in turn, represent the superior forces of an (assumedly) hostile society. That the human infant is physically helpless is perfectly true. But that its encounters with the world are necessarily with hostile forces does not follow. There are powerful but *benign social* forces operating in every known culture to facilitate the growth of the infant. And although the human infant is physically helpless, it is socially powerful, able as Bowlby (1969) has shown, to evoke attachment, i.e., affectionate caretaking, in its parents of both sexes. Recent work in infant psychology has indeed emphasized the hypothesis that the human infant is, to quote Harriet Rheingold (1969), "social by biological origin." Our greatest advance over our nonhuman primate cousins lies precisely in the tremendous increase in the impact of nurturing, affectionate, and moral forces on human behavior. Human infants grow up in an ecology in which moral forces exercise the greatest power over their lives.

Human culture, alas, is also Janus-faced. For reasons no one really understands, the history of civilized humanity has involved the exploitation of many by a few, and within this fact about civilization, it must also be said that men have been the exploiters and women subordinate to men. Note that exploitation, warfare, and the subjugation of women are all practiced in the name of the culture's morality. Among nonliterate peoples, there are cultures which are nonexploitative, and in which the sexes are also correlatively, of equal status. These nonexploitative cultures are few in number; the Arapesh, Zuni, and ancient Kung peoples are among the most famous. Their very existence, however, tells us that exploitativeness is not intrinsic in human nature, but the result of historical forces not yet understood. Their existence also suggests the correctness of Engels's idea that women's and children's exploitation is a subcategory of generalized exploitative relations within a society.

Why it is the men and not the women who are the exploiters and the warriors in societies where exploitation and warfare exist is a related, unsolved historical question. Simone De Beauvoir (1957) specifically raised this question when she disagreed with Engels that division of labor and the development of technology were responsible for women's subjugation. Why, she asked, should not an originally friendly relation between the sexes have survived the development of technology? De Beauvoir's answer once again revealed that, even in so sophisticated a thinker, there was an acceptance of the concept that "domination" of one self over another is basic to humankind. De Beauvoir assumed that men have a need for "transcendence," whereas women need only be "immanent." She based this assumption on the notion that the self develops out of interaction with the "other," and took for granted that this interaction is basically hostile, requiring the male self to · dominate the other (female).

I do not know the answer to De Beauvoir's profound question, but I suggest that her answer, which assumed a male need for "transcendence," was rooted in understandable ignorance of relatively recent work on the psychology of infants. As I suggested earlier, it is when we turn to work on infancy that we find observant psychologists talking not about aggression and domination but about the vicissitudes of an affectionate infant-caretaker interaction. When we follow the anthropologists' lead, which tells us that human beings are above all, uniquely social creatures and look at human infants and their caretakers, we find that an exquisite social interaction is the name of the game.

Let us look at just a brief sample of some of the evidence about the social nature of human infants. Two-month-old babies distinguish faces from inanimate objects, and they look longer at the place where a human being was than at the space left by a missing inanimate object. The three-month-old social smile seems to be a given; there is evidence that blind infants smile at three months, just as do sighted ones. In fact, Eibl-Eibesfeldt (1974) told us he has evidence that blind and deaf thalidomide babies, born without arms to touch their mothers, also smile at three months. And within this evidence for babies' social nature, there are indications that girl infants may have something of an edge over infant boys in their social responses.

Some careful studies have been conducted over the past 15 years of the mother-infant interaction (Moss, 1974). The studies of mothers and their infants began during pregnancy; carefully controlled, reliable observations were made both at home and in the laboratory. Although these studies were not designed to investigate sex differences, and their authors regret this now, infants of both sexes were studied, and some fascinating differences between boy and girl infants emerged. Perhaps the best overall summary of the sex difference I can give is this: girls' socialization is easier; the mother-infant interaction is smoother when mother is dealing with her girl rather than boy infant. For example, at three weeks, boys were crying more, were more irritable, and slept less than girls. At three months, although the male infant still cried more, their mothers attended them less. Mothers "stressed the musculature" of boy babies; they also held them more distant from their own bodies, and they imitated their girl babies more.

It is obvious that two factors must conjoin to produce such results: mothers have different attitudes toward their boy and girl infants, especially in a society which fosters the myth of male superiority, *and* boy and girl babies may bring something just a bit different into the mother-infant interaction. As to the second factor, I am aware that the notion of genetically determined differences between the sexes is not fashionable, especially since the differences are, as our conference notes, used to promote the subjugation of women. But it also seems to me useless to make the mistake of ignoring genetics just because their input has been distorted. I, for example, used to believe that it was impossible in the present climate of women's social inferiority to obtain any meaningful results about genetically determined behavior differences between the sexes, just as it *is* impossible to obtain any meaningful findings on the genetic determination of black-white differences in intelligence. After doing the research for my book, I changed my mind about the analogy between sex differences and black-white differences. The reason is this: in the case of differences between blacks and whites, there are no clearly differentiated gene pools. But when it comes to the difference between the sexes, the difference between having an XX or an XY as the 23rd pair of chromosomes is tremendously powerful. Not that males and females cannot have the same genes, derived from the other 22 pairs of chromosomes. But the difference between having an XX and XY is enormous, that is, having a reproductive system equipped to bear children and nurse them in contrast to a system with a penis and testicles.

Although many studies were not designed to investigate sex differences, there were some most suggestive sex differences between two to three-day-old infants (Korner, 1974). For example, newborn girls suck, mouth, and *smile* more than boys; newborn boys startle more than girls (and they have penis erections). And while the two day-old smile is not a social smile, it is surely possible to consider it a forerunner of the three-month-old social smile, and to at least consider the possibility that girls have some extra sociability endowment than little boys. Another study of newborns also got suggestive and unexpected results (Simner, 1971). Simner was curious as to how newborns respond by crying to the cries of other newborns. This is a phenomenon to which every nursery staff will attest. In a carefully controlled experiment in which he offered newborns the sounds of crying in scrambled wave-lengths, sound of six-month-olds crying, and sounds of newborn cries, the investigator found that the sounds of newborn crying is the most effective stimulus (a finding which suggests an auditory template). And it was newborn girls who were more sensitive to newborn crying than newborn boys. Again, one cannot help imagining that this is some kind of forerunner of women's greater social sensitivity to others. Sensitivity is, of course, a characteristic of women which our social order both fosters and devalues. But because an exploitative social order devalues sensitivity is no reason to avoid the information that women may possess more of it than men if that information turns out to be true.

I now return to my thesis that differences between the sexes are a product of genetics in interaction with the social inequality of women over the affectionate nature of the species. That an exploitative social order injures humanity is not a novel idea: Marx spoke of the alienation which capitalism breeds in its workers. Erich Fromm has developed in detail some of the characteristic distortions which a profit system inflicts upon the people reared in it. In another tradition, Rousseau was struck by the intrinsic psychic injury which civilization inflicts upon its members, the loss of "Pitie'" or sympathy of one human being for another. (This notion of Rousseau's was most influential in Levi-Strauss's thinking.) Freud supposed that all social orders, exploitative and nonexploitative, were built upon the suppression of sexuality—a term he used loosely to include affectionateness and nurturing as well.

My thesis is that the injury which an exploitative society inflicts upon human beings is that it requires them to suppress, renounce, or repress their affectionateness, thus creating a profound internal conflict. An exploitative society requires people to treat other people as if they were things and, as a result, it severely taxes the superego of both sexes. My thesis is also that an exploitative society has a different, although equally severe impact of injury upon the two sexes.

The working model I have developed to understand psychic injury focuses on the superego. The internalized conflict between exploitative and affectionate values catches men and women at different points in their acculturation and along differing routes of internalization of values so that the superegos of men and women tend to operate in different modes. Let me emphasize that the differing modes represent equally developed ethical standards. Specifically, women are more prone to the shame of "loss of love" and to the shame of social inferiority, while men are more prone to guilt for the more frequent transgressions against their own affectionate natures which the exploitative world requires of them.

Let us look more closely at the picture of women first. Little girls, who may already have some kind of edge in sociability over boys, are trained and encouraged in affectionate, nurturing roles. Their gender identity involves only the emulation of their nurturing mothers. But by the time they are two years old, when their gender identity is well established, they discover that affectionateness is not really a useful commodity in an exploitative world. On the contrary, it is a handicap which brings women into "dependent" relationships in which the other, not the self, is the center of the world. Moreover, women, affectionate though they may be, are second-class citizens in the world of power. One consequence is that women have a terrible sense of loss when the others around whom their lives have been built are no longer there to be nurtured. And their sense of loss and helplessness when they lose their nurturing occupation is often intensified because they have no other occupation and often no gainful one. So it is no surprise that women are prone to the shame of loss of love and to the shame of second-class citizenship in the world. It is also a well-established fact that women are two to three times more prone to depression and the hysterias than men. As I have shown in my book *Shame and Guilt in Neurosis* (Lewis, 1971) the affinity between shame experience and both depression and the hysterias is quite close.

It seems to me in keeping with the fact that women do not give up their affectionateness, only devalue it, that the clinical picture in depression and hysteria involves no bizarre distortions of human behavior. The symptoms in both depression and hysteria tend to be common and mundane; as understandable to all of us as depressed mood and/or aches and pains.

The demographic characteristics of depression can also be seen to fit the picture I am describing. Depression cuts across class lines: if anything, there is some slight tendency for depression to be associated with affluence. That depression is a disorder which results from a failure of high ideals of devotion to others was neatly demonstrated in a study by Pauline Bart

(1971). Bart predicted that, on the basis of a strong Jewish tradition of wifely and motherly devotion to the family as a center of their lives, middle-aged Jewish women should be more prone to depression than other ethnic groups in which this particular tradition is not so strong. In an empirical study done in California, Bart confirmed her hypothesis. As she puts it, you don't have to be Jewish to be a Jewish mother (and get depressed) but it helps!

For little boys, the internalized conflict between the affectionate natures and the exploitative demands of the social order in which they are reared assumes a different pattern. Little boys are not encouraged to develop their sociability. On the contrary, by the time their gender identity is established at age two, they are aware that they are expected to be aggressive. Their gender identity itself involves extracting from experience with an opposite-sex caretaker that they are different from her, and more like their more distant fathers. They are expected, before they are six, to "renounce" their affection for their mothers and their identifications with her and to become like their not only distant but more aggressive fathers. As one psychologist (Lynn, 1961) put it, boys have to solve a problem in person-identification, not just take a lesson, the way girls can, in person emulation. At the same time, the culture sets limits on just how aggressive, i.e., just how guilty men may be. No wonder boys, when they grow up, are not only more prone to problems of gender identity but to all manner of obsessional and compulsive disorders, including addictions and sexual perversions! The phenomenological affinity between guilt and obsessive and compulsive states is quite close. Men are also more prone than women to schizophrenia, especially of the paranoid type, in which projection of guilt is an outstanding characteristic.

Schizophrenia, all forms, has entirely different demographic and clinical characteristics. It is strongly associated with class. If you are poor, black, and male, your chances of falling ill of schizophrenia are much greater than if you are rich, white, and female. These facts seem to me to reflect the direct, early attack our exploitative cultural order makes on men's sociability (perhaps going along with men's slightly lesser degree of innate sociability). And the facts go along with the chronically high unemployment rates which make so many men face an apparent failure to achieve success in the world of work. Schizophrenia seems to me to be a response to the demand which our exploitative social order puts more directly upon men than women to function in the world as if aggression and not affectionateness were the stuff of which they are made. The greater guilt which results from men's extensive suppression of their affectionate natures, brings with it the clinical picture of bizarre distortions of gender identity, of the self and of the world which characterizes schizophrenia, i.e., out-and-out social withdrawal into madness.

LITERATURE CITED

BART, P.
    1971. Depression in middle-aged women. *In* Gornick, V., and B. Moran (eds.), Woman in sexist society. New York, New American Library.
BOWLBY, J.
    1969. Attachment and loss. New York, Basic Books.
DE BEAUVOIR, S.
    1957. The second sex. New York, Knopf.
EIBL-EIBESFELDT, I.
    1974. Love and hate. New York, Schocken Books.
KORNER, A.
    1974. Methodological considerations in studying sex differences in the behavioral functioning of newborns. *In* Friedman, R., R. Richart, and R. Van de Wiele (eds.), Sex differences in behavior. New York, Wiley and Sons.
LEWIS, H. B.
    1971. Shame and guilt in neurosis. New York, International Universities Press.
    1976. Psychic war in men and women. New York, New York University Press.

LYNN, D. B.
 1961. Sex role and parental identification. Child Development, vol. 33, pp. 555-564.
MORGAN, E.
 1972. The descent of women. New York, Bantam Books.
MOSS, H.
 1974. Early sex differences and mother-infant interaction. *In* Friedman, R., R. Richart, and R. Van de Wiele (eds.), Sex differences in behavior. New York, Wiley and Sons.
RHEINGOLD, H.
 1969. The social and socializing infant. *In* Goslin, D. (ed.), Handbook of sociabilization theory and research. Chicago, Rand McNally.
SIMNER, M.
 1971. Newborn response to the cry of another infant. Developmental Psychol., vol. 5.

SOCIETY AND GENDER
Eleanor Leacock, Ph.D.
City College, CUNY

One can take one's pick among conflicting generalizations made about women cross-culturally and about the role of women in any specific society; e.g., that "all real authority is vested" in the women of the Iroquois of New York. "The lands, the fields, and their harvest all belong to them. They are the souls of the Councils, the arbiters of peace and of war (Lafitau, 1724)." Or another statement made a century later, that the Iroquois men "regarded women as inferior, the dependent, and the servant of men," and that "from nurture and habit, she actually considered herself to be so" (Morgan, 1954).

Steven Goldberg (1973) predictably chose the second statement. In fact, he liked it so well that he cited it in three separate places in his book. That it was written in the nineteenth century, when the Iroquois lived in single-family houses, and the women were dependent on wage-work done by the men, was of no moment to him. The first statement was written when the Iroquois still retained a measure of political and economic autonomy. Then they lived in the "longhouse," in multifamily collectives. The women owned the land, farmed together, and controlled the stores of vegetables, meat, and other goods. They nominated the sachems who were responsible for intertribal relations, and had the power to recall those who did not represent their views to their satisfaction.

As another example of contrasting reports on women's roles, one can read an account of the Wyandot, a Huron Indian group, written by the nineteenth century anthropologist James Wesley Powell. According to Powell (1880) the Wyandot followed the common native North American practice of separating civil from military matters. Military affairs were decided upon by a council of male warriors who were responsible for fighting, and civil matters were in the hands of clan councils made up of four female household heads and a man of their choosing. The women councillors were responsible for social and economic decisions— allocation of lands, inheritance, marriages, and the like. However, one can read in a contemporary monograph culled from historical accounts of the Huron, that "whatever power women may have had was wielded behind the scenes." Politics was a man's business; the focus of a woman's interest remained within her family and household (Trigger, 1969). Needless to say, Powell's account is not listed in the otherwise full bibliography to be found in this monograph.

And so it goes. A much studied, reported on, and filmed people living today on the borderland between Brazil and Venezuela, the Yanamamo, are characterized in a widely read anthropology textbook (Harris, 1975) as having a style of life that "seems to be entirely dominated by incessant quarreling, raiding, dueling, beating, and killing." The culture is "regarded as among the world's fiercest and most male-centered cultures," the account continues, and "Yanamamo men are as tyrannical with Yanamamo women as Oriental monarchs are with their slaves." In explanation, the author cites increasing population density and struggle over new hunting lands (p. 279).

In a study of another Yanamamo group, however, one reads that these people may have first gained their reputation for fierceness when they fought off a Spanish exploring party in 1758 (Smole, 1976). In that period, Spanish and Portuguese adventurers were ranging throughout the Amazon area searching for slaves. The author of the account worked with a relatively peaceful highland group, and he suggested that the exaggerated fierceness of the lowland Yanamamo is not typical, but may have been developed for self-protection. In the village he studied, elder women, like elder men, are highly respected. When collective decisions are made, mature women "often speak up, loudly, to express their views." Younger men, like younger women, "have little influence" (p. 70). There were two widowed matriarchs in the village where he worked. Each was an "old, highly respected woman whose needs are met fully by her own children, her sons- and daughters-in-law, her grand-children, and her nieces and nephews. Much concern is shown for such a woman's comfort and well-being" (p. 75).

Skipping to another major world area, one can read of "the traditional ideal of male domination characteristic of most African societies" (Le Vine, 1966). Or one can read that in most of the monarchical systems of traditional Africa, there were "either one or two women of the highest rank who participate in the exercise of power and who occupy a position on a par with that of the king or complementary to it" (Lebeuf, 1971). According to Lebeuf, women's and men's positions were complementary throughout the various social ranks of African society. Women formed groups for "the purpose of carrying out their various activities," and these could become "powerful organizations." An example of how such groups have functioned even in recent times is given in an account of the Anaguta people of Nigeria. When news spread among the women that a new ruling might cut off the income they made from the sale of firewood, they

> marched down from the hills and assembled before the courthouse in a silent, formidable, and dense mass, unnerving the chief and council, all of whom made speeches pledging sympathy; and similar demonstrations took place when it was rumored that women were going to be taxed in the Northern Region (Diamond, 1970).*

A recurrent theme in contemporary anthropological literature is that men's activities are always in the public and important sphere, while women's concerns are limited to the private, familial, and subsidiary sphere. Le Vine (1966, p. 187), who

---

*This event took place in the early 1960 s. The famous women's demonstrations in Nigeria took place in the 1920s, in one of which 30 women were shot.

wrote of traditional male domination in African society, stated that "women contribute very heavily to the basic economy, but male activities are much more prestigeful." By contrast, Lebeuf (1971, p. 114) wrote, "neither the division of labour nor the nature of tasks accomplished implies any superiority of the one over the other."

Another anthropologist, Sudarkasa (1976), stressed the complementarity of traditional African social-economic organization. She described women's organizations as vehicles of cooperation and collaboration with men, as well as a means of defending women's interests when necessary:

> Whereas from the perspective of the twentieth century, the public sphere of a society can be defined . . . in terms of political and economic activities that extend "beyond the localized family unit," when one looks at the preindustrial, precapitalist, and precolonial world, it becomes obvious that many such political and economic activities were in fact embedded, albeit not exclusively, in domestic units . . .
>
> The existence of female chiefs and of other female leaders in the public sphere should not be interpreted as evidence of their achieving status by "entering the world of men." This formulation misses the essential point that the "public sphere" in most West African societies was not conceptualized as "the world of men." Rather it was one in which both sexes were recognized as having important roles to play.

The point Sudarkasa made for West Africa can be broadly applied. The structure and images of contemporary western society are often projected onto other cultures uncritically when women's roles are being discussed, and historical changes that took place with the spread of colonialism and imperialism are ignored. The sheer lack of information on the activities of women and decisions made by them has encouraged this ethnocentrism. However, evidence now being gathered indicates that "male dominance" is not a human universal, as is commonly argued; that in egalitarian societies the division of labor by sex has led to complementarity and not female subservience; and that women lost their equal status when they lost control over the products of their work (Leacock, 1977).

Today, the age-old *practical* basis for a sex division of labor according to reproductive roles and responsibilities has all but vanished. Assertions of past inferiority for women should therefore be irrelevant to present and future developments. It is all the more interesting, therefore, that a new statement of women's "natural" role "in the home" is being put forth with prestigious academic backing and considerable media attention. In the much publicized book, *Sociobiology, the New Synthesis,* the Harvard professor Edward O. Wilson (1975a) wrote:

> The building block of nearly all human societies is the nuclear family. The populace of an American industrial city, no less than a band of hunter-gatherers in the Australian desert, is organized around this unit. . . During the day the women and children remain in the residential area while the men forage for game or its symbolic equivalent in the form of barter and money.

In the New York Times Magazine (1975b), Wilson stated his view of the matter unequivocally. "In hunter-gatherer societies men hunt and women stay home. This strong bias persists in most agricultural and industrial societies, and on that ground alone appears to have a genetic origin." Wilson believed a sex division of labor will therefore exist even in the most democratic of future societies, although he hastened to add that this should not justify job discrimination on the basis of sex.

The book Wilson draws on for information about gathering hunting societies (Lee and DeVore, 1968) includes the well established fact that in most of them women furnished as much if not more of the basic food as did men. Women dug roots, gathered seeds, fruits and nuts, and/or collected shellfish— staple foods on which their bands heavily depended. They certainly could not stay in camp to do this. They commonly foraged in groups, often miles from home base, and distributed the food, not just to their nuclear families, but to networks of kin. Much the same held true for stone-tool using horticultural societies, such as the Iroquois, where the women worked the fields in groups and also foraged widely for wild vegetable foods. (Women also did some hunting in such societies according to varying traditions, or individual need or desire. Generally speaking, taboos against women hunting are found in societies with rank or class differences.)

When it comes to our own historical past, Wilson should be reminded that the word "family" comes, not from the revered nuclear unit of nineteenth century Europe, but from the household of wife, concubines, slaves, and children over which a Roman patrician had rights of life and death. He should also consider that all but upper-class women were expected to work in medieval Europe and that nunneries were sometimes virtual business enterprises (Lasch, 1973), that the factory system was built both in Europe and in New England largely upon the labor of women as well as children; and that slave women knew nothing of model Victorian home life. "I am above eighty years old," Sojourner Truth said to a meeting on women's rights in 1867, "it is about time for me to be going."

> I have been forty years a slave and forty years free, and would be here forty years more to have equal rights for all. I suppose I am kept here because something remains for me to do; I suppose I am yet to help break the chain. I have done a great deal of work; as much as a man, but did not get so much pay. I used to work in the field and bind grain, keeping up with the cradler; but men doing no more got twice as much pay... We do as much, we eat as much, we want as much (in Lerner, 1973).

In short, for a large proportion of women in Western history, "in the home" has always referred to variations on a common theme: non-paid service work in it and lower-paid work outside it.

At the same time as much-needed analysis of the different family arrangements, indicated by the foregoing, is getting underway, the entire question of historical changes in women's position is being glossed over by educational materials on "sociobiology." For example, a publicity piece for the film, "Sociobiology: Doing What Comes Naturally[1], states in part:

> Harvard University biologist Robert Trivers speaks about the possibility of sex-determined behavior. Despite the assertions of the women's liberation movement, Dr. Trivers feels that natural selection has been working for centuries to develop emotional dispositions to match the male's natural physical freedom and the female's more vulnerable, childbearing nature.
>
> Anthropologist Irven DeVore discusses the competitive drive for status among any species and the more probable survival of the genes of such dominant individuals. . .
>
> "It's time we started viewing ourselves as having biological, genetic, and natural components to our behavior. And that we should start setting up a physical and social world which matches those tendencies. . ."

This film, like sociobiology itself, skips over the course of human history and ignores the profound transition from egalitarianism to exploitative and hierarchical organization. It treats women purely as child-bearers and ignores them as workers. Men do not fare much better. The images flash back and forth between fighting male baboons and discussions of sex relations among college students, emphasizing the theme of innate male aggressiveness and competition over "passive" females. Although Wilson has disclaimed responsibility for this particular film, he has not done so for the Nova film which says the same thing in a scientific fashion.

[1] Document Associates, Inc., 880 3d. Ave., N.Y.

Here we move into another area in which contradictory statements about sex roles abound, for there is scarcely any limit to the variety of reproductive relations and arrangements that can be found in the animal world. One can pick and choose, and the choices in this film, as in other media material, suggests that the problems of rampant profit-seeking and of war with which we are beset follow from our animal nature, and are rooted in male competition for females. Kolata (1976) stated that 1) Baboon societies vary according to environmental conditions and do not offer the clear-cut example of male dominance suggested by the film. 2) The primary behavioral characteristic of monkeys generally is *sociality*. Aggression (as variously defined) is only one form, and usually a minor form, of social behavior, and females do not necessarily choose "aggressive" or "dominant" males more than others. The statement attributed to DeVore above is not borne out. 3) In any case, our closest relatives are not baboons, but the apes. Among the well-studied chimpanzees, males do not compete for females (Jolly and Plog, 1976). 4) Humans evolved as gatherer-hunters, and from what we know about foraging society, aggression was frowned upon and avoided. The valued attributes were the skills—social, manual, artistic, intellectual—and they were valued in both women and men. Authors who gave a rounded picture of life among huntergatherers are Turnbull (1962) and Washburn and Amanta (1940).

Yes, one can take one's pick among conflicting generalizations. This has been true since the times of John Locke and Thomas Hobbes. Locke, defender of democratic forms, stressed human cooperativeness, and cited as an example the generosity of native Americans who were still free, living without rulers apart from centers of colonial conflict; while Hobbes, defender of a strong monarchy, argued that the competitiveness of his times was innate. What we understand about ourselves is crucial. Today the humanistic goal of a peaceful and cooperative world has become an urgent need if we are to survive as a species. Generalizations about women are, in effect, generalizations about men and about human society in general. It is important to pick right.

LITERATURE CITED

DIAMOND, STANLEY
1970. The Anaguta of Nigeria: Suburban primitives. *In* Steward, Julian H., ed., Contemporary change in traditional societies, vol. I: Introduction and African tribes. Urbana, University of Illinois Press, p. 476.

GOLDBERG, STEVEN
1973. The inevitability of patriarchy. New York, Wm. Morrow, pp. 40, 58, 241.

HARRIS, MARVIN
1975. Culture, people, nature. New York, Crowell, pp. 276, 279, 399.

JOLLY, CLIFFORD J., and FRED PLOG
1976. Physical anthropology and archaeology. New York, Alfred A. Knopf, p. 61.

KOLATA, GINA BARI
1976. Primate behavior: sex and the dominant male. Science, vol. 191 (January 9), pp. 55-56.

LAFITAU, JOSEPH F.
1724. Moeurs des sauvages Ameriquaines, comparees aux moeurs des premiers temps. Cited in Brown, Judith K., Iroquois women: An ethnohistoric note. *In* Reiter, Rayna R., Toward an anthropology of women, New York, Monthly Review Press, p. 238.

LASCH, CHRISTOPHER
1973. Better than to burn. *In* McNamara, Joan, and Suzanne F. Wemple, an historical note on marriage. The Columbia Forum (Fall, 1973). Sanctity and Power: Medieval Women, *In* Bridenthal and Koonz (eds.), Becoming visible, women in European History. Boston: Houghton Mifflin, 1977.

LEACOCK, ELEANOR
1977. Women in egalitarian societies. *In* Bridenthal, Renate, and Claudia Koonz (eds.), Becoming visible, women in

European history. Boston, Houghton Mifflin.
LEBEUF, ANNIE M.D.
   1971. The role of women in the political organization of
   African societies. *In* Paulme, Denise, (ed.), Women of
   tropical Africa. Berkeley, University of California Press,
   pp. 97, 113, 114.
LEE, RICHARD B., and IRVEN DEVORE (eds.)
   1968. Man the hunter. Chicago, Aldine, pp. 41-43.
LERNER, GERDA
   1973. Black women in White America. New York, Random
   House, p. 570.
LEVINE, ROBERT A.
   1966. Sex roles and economic change in Africa. Ethnology,
   vol. V, no. 2, pp. 187, 192.
MORGAN, LEWIS HENRY
   1954. League of the Ho-De-No-Sau-Nee or Iroquois, vol. I.
   New Haven, Human Relations Area Files, p. 315.
POWELL, J. W.
   1880. Wyandot government: A short study of tribal society.
   In Annual Report of the Bureau of American
   Ethnology, vol. I.
SMOLE, WILLIAM J.
   1976. The Yanamamo Indians, a cultural geography. Austin,
   Univ. of Texas Press, pp. 15, 70, 75.
SUDARKASA, NIARA
   1976. Female employment and family organization in West
   Africa. *In* McGuigan, Dorothy G., (ed.), New research
   on women and sex roles. Ann Arbor, Center for
   Continuing Education of Women, University of
   Michigan, pp. 50, 53-54.
TRIGGER, BRUCE G.
   1969. The Huron, farmers of the North. New York, Holt,
   Rinehart and Winston, p. 74.
TURNBULL, COLIN
   1962. The forest people. Garden City, New York, Doubleday.

WASHBURN, HELUIZ ,and CHANDLER, and AMANTA
1940. Land of the good shadows. New York, John Day.
WILSON, EDWARD O.
1975a. Sociobiology, the new synthesis. Cambridge, Mass.,
Harvard University Press, p. 553.
1975b. Human decency is animal. New York Times Magazine.
October 12.

EPILOGUE

The thoughtful papers presented here show that for the most part women reject the concept of genetic destiny and hereditarianism. They also show that there is not yet agreement on the significance of the concept nor on how it expresses itself in our professional and scientific thinking. We are concerned that discrimination and oppression are taught in our schools, and preached in our media and practiced daily in all phases of our society. These activities are justified by many educators, philosophers, and scientists who believe that genes determine our lot in life. Scholars and scientists who promote hereditarianism are likely to continue doing so so long as the society in which they work supports them and rewards them for their antihuman activities.

One of the most widespread expressions of hereditarianism is the concept of "instinctive" behavior. Behavioral theories such as ethology and sociobiology are based on the idea that genes program certain behavior; these are called instincts. Many women can see that racism is perpetuated by the myth of genetic programming or instincts, but its use to discriminate against women is not clear; let's look at "maternal instinct." The consequences of this myth is that only a woman should care for babies; that it is good for a woman to be kept in this "noble" activity. Satisfying this "instinct" also prevents her from being available for other kinds of work and keeps her in her "place" in society. The same logic is applied to the male "instinct" for dominance and aggression.

Many women believe that the "instinctive" expressions of reproduction and physiology cannot be denied. Should a woman not "act them out" through sexual intercourse or having children, she will become ill or incapacitated in some way as a human being. There is no scientific evidence to support these myths. Physiological health does not depend on having sexual intercourse or bearing children. The belief that they *are* necessary to prove manliness or womanliness can lead to psychological ill-health. This idea can cause frustration, anxiety and hostility, and can affect all interpersonal relationships. Scientists who pronounce that reproductive behavior is an inborn drive, need or instinct add to the pressure that society produces on women and men.

The belief that instinct, natural selection or genes has programmed our destiny leads to a concern that this inborn need cannot be satisfied in a society that exploits women. Thus, the slogan for "sexual freedom" has been one of the earliest focuses of the struggle for equal rights. This was in direct response to the exploitative nature of the relationship between women and men and between different races in which sexual behavior was used as a weapon of dominance and terror.

The freedom to engage in any type of interpersonal activity in a nonexploitative, nondemeaning relationship is indeed a significant goal to fight for. However, we suggest that the basis for that freedom does not lie in the inalienable right of people to satisfy their instincts but in the human right to be free of any oppression and of any violation of their societal dignity and personal integrity.

Orthodox Freudian psychology, by teaching that psychosexual development follows a predetermined course, has also added an aura of "scientific" respectability to another chain that hampers people in their struggle to achieve such freedom. Ethologists and Freudians have long agreed in their writings and research about the instinctive basis of behavior.

More recently, Wilson has proposed that all the behavioral sciences (anthropology, psychology, sociology) have not gone far enough in incorporating evolutionary theory and urges a rethinking of basic principles so that they can be made compatible with his theoretical position. He says that all behavior in all animals, including people, and particularly social behavior is organized to do one thing primarily: to pass the individual's genes on to the next generation. (This oversimplification of evolution leads to looking at the world through a telescope as though it were a microscope.) In the sociobiology system of thought, the relationship between human parents and their children becomes a source of conflict if one assumes a limited energy system. On the one hand, the parent wants to survive as an individual and to have more offspring. On the other hand, the children "have" half of each parent's genes and should be taken care of by the parent so that they might be able to have offspring who will also carry some of the parent's genes. Similarly, there is a genetically determined conflict between women and men: the more children one has by more people the more likely one's genes will be spread around. However, our societal values place a high premium on loyalty to one spouse but this is applied more to women than to men--the "double standard."

Many scientists are exposing the pseudoscience on which Wilson's theories are built. Genes are biochemical systems and are expressed in the development of other biochemical systems. Even this expression of the genes is a function of the particular physical and chemical characteristics of the cell or tissues in which they function. Between the functional level of these molecules, enzymes and other minute biochemical living systems, and behavior and society, there are many factors that determine how these genes will act. It is not that genes determine us, but that we, through our societal behavior, determine how genes will function.

First, societal processes determine whether one or another woman will become pregnant and by whom. Then, the conditions of her pregnancy, whether the child will be born, how that child will develop, and what that child will do with its life are further determined by societal processes.

For example, particular genetic configurations are important for the function of glands, such as the thyroid. However, healthy thyroid function is affected by diet. When the thyroid gland does not function normally this affects behavior. One of the diseases of dysfunction of the thyroid is goiter. In a society where there is iodine in the diet, children are less likely to develop goiters than are children living in a place where there is a deficiency of iodine. In both places, not *all* will be free of goiter. In both places, the function of the biochemical systems will vary and be differentially affected by the diet. There are many possible combinations of biochemical characteristics (genes). We do not know which combinations exist in any one individual or in any population of individuals until they have had an opportunity to function and express themselves under a variety of specific conditions and situations.

The function of genes in reproduction is similarly complex. Most sexually reproducing organisms, including humans, show four aspects of genetic expression: chromosomal; gonadal; hormonal; gamete production (see papers by Probber and Ehrman, and by Briscoe). Only human beings are assigned "gender," that is, roles in society that differ for women and men. These roles may or may not be related to chromosmal, gonadal, hormonal or gametic "sex."

It is often advantageous for those in power to attempt to set people against one other because of race, sex, or ethnic background. Hereditarianism is used to justify such discrimination. As long as people are kept apart on the basis of belief in hereditarianism and genetic destiny they cannot overcome their oppression. Together, women, Blacks, Hispanics, Native-Americans and all oppressed minorities can expose the myths used to defeat them.

Women and men who understand this will see to it that the conditions in which people are born will give each person the best chance for the healthiest development. In that setting our understanding of gender, or differential societal roles for women and men, will be very different, and oppression and exploitation will be eliminated.

This is our first attempt to discuss these problems in a conference. We plan to continue conferences to explore ideas and ways in which women can turn science into a force for their liberation.

Ethel Tobach and Betty Rosoff